Particle Technology Forum 2022

Held at the 2022 AIChE Annual Meeting

Phoenix, Arizona
13-18 November 2022

ISBN: 978-1-7138-7916-9

Printed from e-media with permission by:

Curran Associates, Inc.
57 Morehouse Lane
Red Hook, NY 12571

Some format issues inherent in the e-media version may also appear in this print version.

Printed with permission by Curran Associates, Inc. (2023)

For permission requests, please contact AIChE
at the address below.

AIChE
120 Wall Street, FL 23
New York, NY 10005-4020

Phone: (800) 242-4363
Fax: (203) 775-5177

www.aiche.org

Additional copies of this publication are available from:

Curran Associates, Inc.
57 Morehouse Lane
Red Hook, NY 12571 USA
Phone: 845-758-0400
Fax: 845-758-2633
Email: curran@proceedings.com
Web: www.proceedings.com

Particle Technology Forum 2022

Held at the 2022 AIChE Annual Meeting

Phoenix, Arizona
13-18 November 2022

AIChE
The Global Home of Chemical Engineers

TABLE OF CONTENTS

ENGINEERED PARTICLES AND NANOSTRUCTURED PARTICULATE SYSTEMS CHARACTERIZATION

FLUIDIZATION: EXPERIMENTAL INVESTIGATION AND MODELING OF FLUIDIZATION PROCESSES

FLUIDIZATION: FUNDAMENTALS

FLUIDIZATION: MODELING

FUNCTIONAL NANOPARTICLES AND NANOCOMPOSITES

NANOSTRUCTURED MATERIALS FOR PHARMACEUTICAL APPLICATIONS

NOVEL NANOPARTICLES AND NANOSTRUCTURED CATALYSIS FOR ENERGY AND ENVIRONMENTAL APPLICATIONS

NOVEL NANOPARTICLES AND NANOSTRUCTURED MATERIALS FOR PHARMACEUTICALS AND MEDICAL APPLICATIONS

PARTICLE AGGLOMERATION AND GRANULATION PROCESSES

PARTICLE BREAKAGE AND COMMINUTION PROCESSES

PARTICLE TECHNOLOGY FORUM AWARD PRESENTATIONS AND BARON AWARD LECTURE (INVITED TALKS)

PARTICULATE SYSTEMS: DYNAMICS AND MODELING: APPLICATIONS

PARTICULATE SYSTEMS: SOLIDS HANDLING, PROCESSING, CONVEYING, SEPARATION, AND HEAT TRANSFER

PHARMACEUTICAL POWDER AND PARTICULATE SYSTEMS

POSTER SESSION: PARTICLE TECHNOLOGY FORUM

POWDER AND PARTICULATE CHARACTERIZATION AND MEASUREMENT

PARTICULATE SYSTEMS: DYNAMICS AND MODELING: DISCRETE/CONTINUUM MODELS

PARTICULATE MIXING AND SEGREGATION

PARTICLE TECHNOLOGY: HONORING JOHN CARSON

PARTICLE TECHNOLOGY IN PRODUCT DESIGN AND MANUFACTURING

CRYSTALLIZATION IN PROCESS DEVELOPMENT

SESSION HONORING LIFETIME ACHIEVEMENT OF MADHAVA SYAMLAL

407f Effect of Monosaccharide and Polysaccharide Excipients on APAP Tablets

Sheena Reeves, Nigel Brooks Jr.

Most pharmaceutical excipients serve as binding agents in tablet formation. As an inactive pharmaceutical ingredient, an excipient should not interfere with the drug absorption into the body. Synthetic materials, such as polyethylene glycol (PEG), are well-known for their binding attributes in drugs. However, synthetic binders have become more complex in recent years which have resulted in higher levels of toxicity and a reduced solubility of the active pharmaceutical ingredient (API). Due to these factors, researchers are investigating inert, natural binding materials such as monosaccharides and polysaccharides. This research sought to investigate the effect of replacing PEG as an excipient in paracetamol tablets with monosaccharides: d-allulose and d-sorbitol; and polysaccharides: tamarind gum, pectin, guar gum, and xanthan gum. The tablets were formed using a binder dropping method into a 3D printed tablet mold. Analysis reviewed the effect of excipient selection on the crystal structure of the tablet using XRD, effect on thermal properties using DSC, and effect on the chemical structure using FTIR. Overall, excipient selection resulted in no significant changes in the APAP crystal structure which would signal the formation of a different compound. FTIR analysis showed that polysaccharide excipients do not affect the structure of APAP, while the sweeteners interacted with the -OH functional group. The synthetic excipient, PEG, had the greatest effect on the structure of APAP and was the only excipient producing new functional group peaks. All excipients reduced the maximum melting temperature of the APAP tablet with the polysaccharides having the smallest change. D-sorbitol provided the highest reduction in melting peak and onset temperatures.

407b Crystal Growth Impedance from Boundary Layer Transport, Conformational Interconversion, and Dimerization Kinetics

Armin Shayesteh Zadeh, Baron Peters

Some crystal grow by exclusively incorporating one conformer or by exclusively incorporating pre-formed dimers. If the conformational interconversion rate (or the dimerization rate) is slow compared to the growth unit incorporation rate, then the kinetics of conformational interconversion (or dimerization) may impede the growth kinetics. We model these effects to quantify the degree of crystal growth impedance in mixed-suspension-mixed-product-removal reactors and for transport through a boundary layer to the growing crystals. Our analysis reveals simple expressions and dimensionless variables that can be used to anticipate when and how severely the kinetics of conformer interconversion and/or dimer formation will impact the growth kinetics. Finally, we use the model predictions to assess the impact of conformer interconversion kinetics in earlier experimental and computational reports.

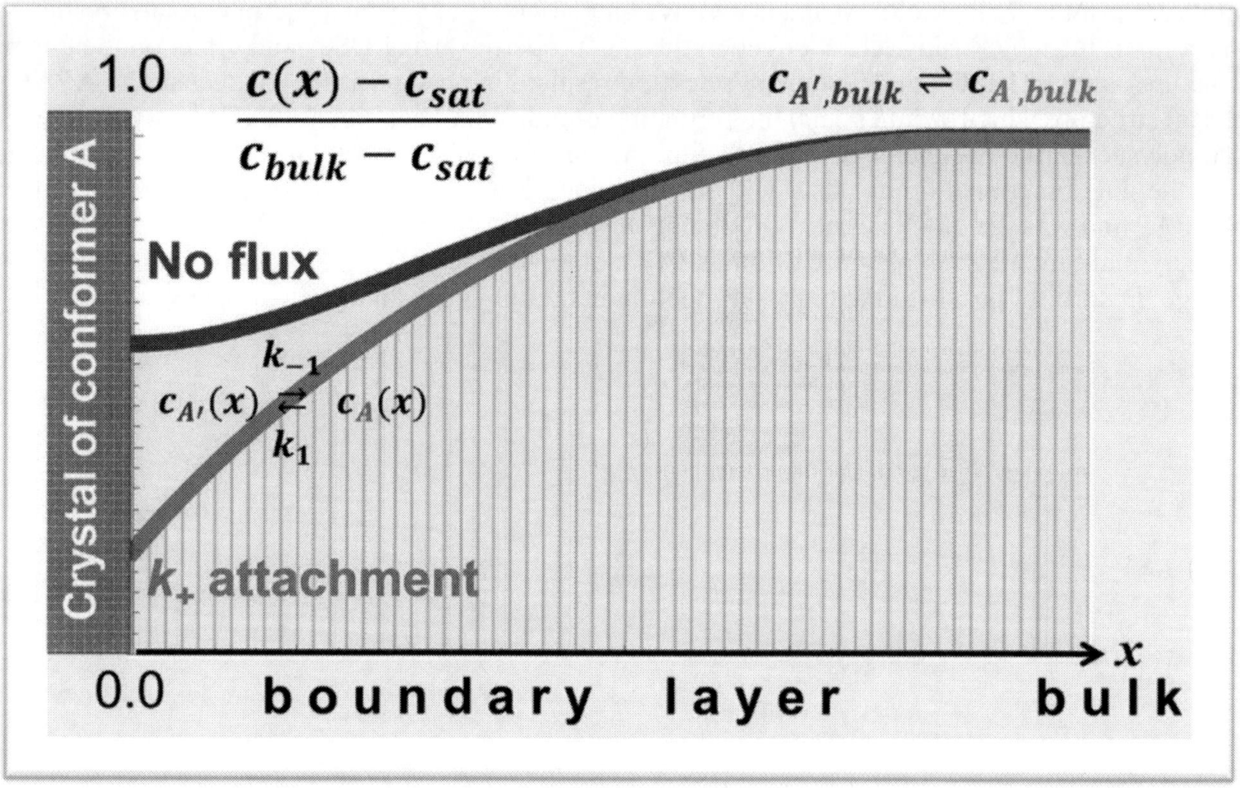

407c Direct Ink Writing of Tungsten and Tungsten Alloys for Dense Complex Shapes

Hailey Loehde-Woolard, Elena Napoletano, Davis R. Conklin, Bergen Evans, James Smay, Alan Weimer

Tungsten (W) is a refractory metal that exhibits a unique combination of high melting point, strength, ductility, thermal conductivity, and corrosion resistance. As such, it is integral for use in extremely high temperature applications such as nuclear thermal propulsion and shielding elements of solar probes. However, due to the high melting point (3422°C), W is extremely difficult to melt and cast into various complex shapes. Most W based alloys are made through powder metallurgy, pressing and sintering W powder with the addition of transition metal sintering aids. This limits the complexity of the final fabricated part, limiting the ability to produce intricate components for the desired applications. Current state of the art additive manufacturing techniques for W and W alloys are powder bed fusion and binder jet printing. Both techniques are highly dependent on process conditions, which can introduce flaws such as stress gradients, porosity, and non-uniform shrinkage in the final part.

In this work we propose the use of direct ink writing (DIW) of W and W alloys to fabricate complex shaped components with near full density, high yield strength, and low surface roughness. Further, we utilize atomic layer deposition (ALD) to deposit sintering aids and dopants at low loading directly on the particle surface. A tungsten-based colloidal ink was formulated with an optimized solids fraction volume loading. The ink was then continuously extruded through a nozzle which enables bottom-up fabrication of the desired geometry, such as cylinders or lattices for testing. A design of experiments (DOE) was employed to optimize printing parameters in order to mitigate surface roughness and maintain print uniformity. The green bodies were dried and sintered at 1600°C in a low hydrogen atmosphere. Surface profilometry was used to obtain the sintered surface roughness and evaluate the DOE. The optimized printing parameters were employed to manufacture samples with various sintering aids and dopants, added through conventional mixing or ALD on the powder prior to ink formation. A secondary DOE was employed to evaluate the parameters that affect the final print density. The systematic optimization of the colloidal ink resulted in parts that demonstrated both densification and stability in complex geometry. Addition of the sintering aids promoted densification further at the lower sintering temperature and ALD was shown to be a successful loading method. This work introduces the use of ALD-functionalized W in colloidal gels for the purpose of DIW complex shapes.

407d Adapting Cold Spray Technology to Polymer Powders

Tristan Bacha, David A. Brennan, Anuraag Gangineri Padmanaban, Jae-Hwang Lee, Francis M. Haas, Joseph F. Stanzione III

Adapting cold spray technology to polymeric materials

Cold spray processing has proven to be an effective method of applying coatings, repairing structures, and high-rate additive manufacturing.[1, 2] In the cold spray process, micron-sized powders in the solid state are accelerated with a supersonic stream of gas and impacted on a substrate. At ideal material-specific processing conditions, powders will deposit onto a substrate and form a coating due to extensive plastic deformation between the particle and substrate. This process has been shown to produce hard, dense, and efficient coatings from metal powder feedstocks over the past several decades.[1] Recently, interest in applying the technology to polymer powders has been expressed in the literature. [3-7] Although a promising technology, polymer powders have proven to be challenging to spray with reports generally struggling to achieve efficient sprays and high deposition quality. The macromolecular structure of polymers and their thermophysical properties require re-thinking the cold spray process from concepts that were developed from metal particle sprays. Our work investigates how these fundamental polymer properties affect cold sprayed polymers through a combined approach including multiphysics modeling, highly controlled single-particle impact experiments, and the analysis of cold sprayed polymer powders. We highlight some of the critical changes that arise in the process (heat transfer characteristics, polymer chain length, and thermophysical properties) and how the process must be adapted to achieve efficient and quality polymer deposits. Ultimately, a foundational understanding of the process will enable a versatile method of rapidly repairing, building, and coating a range of materials with polymers.

1. Moridi, A.; Hassani-Gangaraj, S. M.; Guagliano, M.; Dao, M., Cold spray coating: review of material systems and future perspectives. *Surface Engineering* **2014,** *30* (6), 369-395.
2. Raoelison, R. N.; Verdy, C.; Liao, H., Cold gas dynamic spray additive manufacturing today: Deposit possibilities, technological solutions and viable applications. *Materials andamp; Design* **2017,** *133*, 266-287.
3. Bernard, C. A.; Takana, H.; Diguet, G.; Ravi, K.; Lame, O.; Ogawa, K.; Cavaillé, J. Y., Thermal Gradient of In-Flight Polymer Particles During Cold Spraying. *Journal of Materials Processing Technology* **2020,** *286*, 116805.
4. Bush, T. B.; Khalkhali, Z.; Champagne, V.; Schmidt, D. P.; Rothstein, J. P., Optimization of Cold Spray Deposition of High-Density Polyethylene Powders. *Journal of Thermal Spray Technology* **2017,** *26* (7), 1548-1564.
5. Khalkhali, Z.; Rothstein, J. P., Characterization of the Cold Spray Deposition of a Wide Variety of Polymeric Powders. *Surface and Coatings Technology* **2020,** *383*, 125251.
6. Ravi, K.; Deplancke, T.; Ogawa, K.; Cavaillé, J.-Y.; Lame, O., Understanding Deposition Mechanism in Cold Sprayed Ultra High Molecular Weight Polyethylene Coatings on Metals by Isolated Particle Deposition Method. *Additive Manufacturing* **2018,** *21*, 191-200.
7. Xu, Y.; Hutchings, I. M., Cold Spray Deposition of Thermoplastic Powder. *Surface and Coatings Technology* **2006,** *201* (6), 3044-3050.

407e 3D-Printing Hybrid Electrolyte Structures for Dendrite Suppressions

Nathan Fonseca

Current lab-scale production of Solid-State Electrolytes (SSE) for Lithium-ion batteries is limited to planar geometries and random porosities. Although SSE has shown improvements via advanced manufacturing techniques, they still face dendrite propagation issues. Additive manufacturing (or 3D printing), known for its rapid prototyping and design flexibility of complex structures, allows the materials and structures to be tuned to achieve a better performance battery. Moreover, different hierarchies (e.g., patternings, layers, pillars) can be incorporated to block dendrite growth and prevent further degradation by eliminating battery safety hazard effects. This work focuses on finding an ideal strategy for 3D printing $Li_{6.5}La_3Zr_{1.5}Ta_{0.5}O_{12}$ (LLZTO) ceramic electrolytes with unique structures for high energy density and high safety characteristics to enable greater resilience to extreme conditions. We will include the synthesis of the particles and electrolytes, design of 3D printing procedures, characterization of 3D printed structures, and performance of energy systems compared to the state-of-the-art

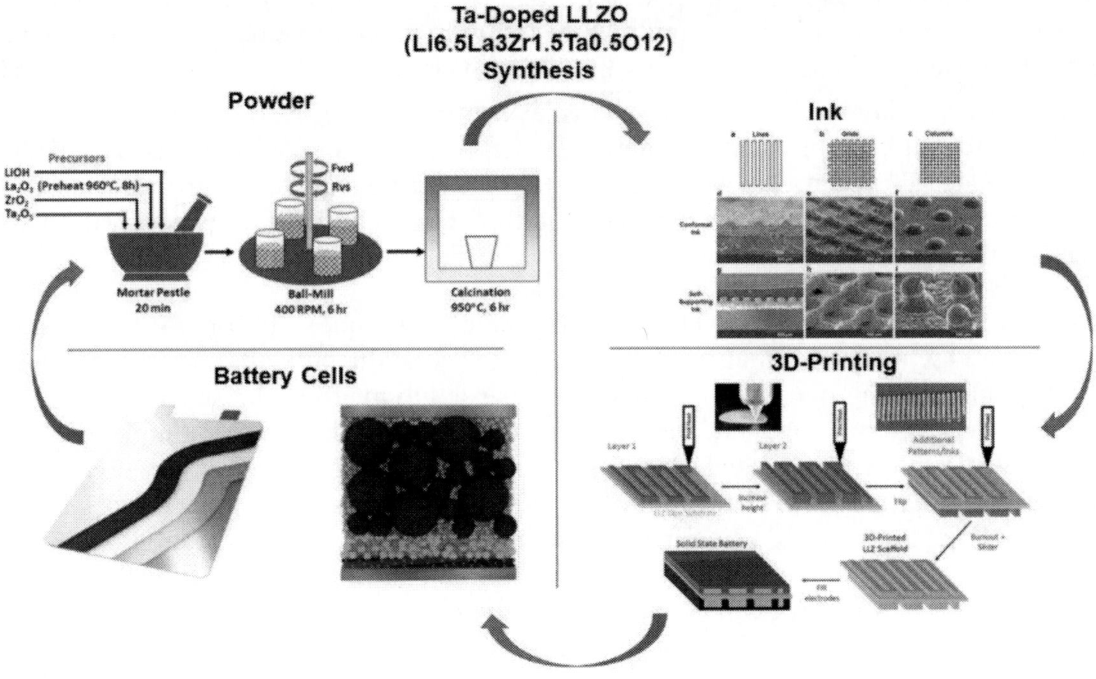

334c Synthesis and Characterization of Aluminum Boride Particles Coated with Plasma Nanofilms for Energetic Applications

Prawal P. K. Agarwal, Themis Matsoukas

Aluminum (Al, 31 kJ/g) and boron (B, 58 kJ/g) based energetic materials oxidize exothermically to produce light, heat, and thrust, which find applications in solid fuels, propellants, and pyrotechnics. The primary inhibiting factor in extracting the stored chemical energy is the native oxide layer present on their surfaces, which occupies a significant mass fraction of the nano and submicron particles, acts as a diffusion barrier to the oxidizer, and does not contribute to the energy. In order to improve the reactivity and energy release, the oxide layer must be removed or chemically converted into materials that promote combustion and enhance the amount of energy released. In this study, we anneal a mechanical mixture of Al and B to synthesize submicron-sized aluminum boride. We optimize the parameters such as the composition of the mixture, time, and temperature through experiments. The advantages associated with Al boride are that the B oxide present in the particles can be reduced to B by metallic Al through an exothermic redox reaction during oxidation. This increases the B content of the material at the expense of less energetic Al, resulting in enhanced energy release. The boride is then treated by plasma-enhanced chemical vapor deposition (PECVD) of perfluorodecalin ($C_{10}F_{18}$) to produce a thin surface layer of CFx species. This layer provides passivation against air and humidity and contributes to combustion via the formation of metal fluorides which contribute to the energy content of the material. Effectively we are converting the native oxides into energetics. Several experiments are performed to find the ideal coating thickness according to its passivation ability and energy release during oxidation.

We characterize the material by a variety of experimental techniques. Its composition is studied by X-ray diffraction and X-ray photoelectron spectroscopy to probe the bulk and the surface, respectively. Particle size analysis using dynamic light scattering is used to measure the size of the synthesized particles. The shape, morphology, and elemental distribution are studied using high-resolution transmission electron microscopy and scanning transmission electron microscopy combined with energy dispersive spectroscopy. The extent of oxidation and corresponding heat release of the synthesized material is measured using thermogravimetric analysis and differential scanning calorimetry. The combustion heat is measured using bomb calorimetry. The core-shell structures formed show great promise as energetic materials with enhanced heat release and more extended storage stability.

135a A Structured Bubbling Phenomenon in Vertically Vibrated Binary Gas–Fluidized Beds

Jagan Mohan Sanghishetty, Christopher Spitler, Qiang Guo, D. R. Nagaraj, Raymond Farinato, Chris Boyce

Granular flows are ubiquitous both in natural and industrial processes such as pharmaceuticals production[1], mining and food–processing. A vital hydrodynamic flow regime inside granular fluidized beds is the free–bubbling regime which exhibits enhanced heat transfer characteristics. With the mathematically chaotic motion of these rising bubbles, the system's characteristics are unpredictable. Recent studies[2, 3] have shown that this chaos can be suppressed by oscillating the gas flow rate or vertically vibrating the fluidized beds. In particular, vertical vibration at the resonant frequency generates periodic, triangular, structured array of bubbles, across various system dimensions and particle sizes[4]. Extending this phenomenon further to the mixtures, here we demonstrate that this dynamically structured bubble configuration persists in binary mixtures composed of different density ratios, different particle size ratios and across different system sizes. Discrete Element Method (DEM) simulations accurately capture the bubble structuring for various initial particle arrangements and also facilitate the quantification of mixing characteristics. The optically imaged experimental data is compared with simulation predictions.

References:

- Muzzio, F.J.; Shinbrot, T.; Glasser, B. J. Powder technology in the pharmaceutical industry: The need to catch up fast, *Powder Technology* **2002**, 124 (1), 1–7.
- Coppens, M. -O.; van Ommen, J. R. Structuring chaotic fluidized beds, *Chemical Engineering Journal* **2003**, 96, 117–124.
- Wu, K.; de Martín, L.; Coppens, M.-O. Pattern Formation in Pulsed Gas-Solid Fluidized Beds – The Role of Granular Solid Mechanics. *Chemical Engineering Journal* **2017**, *329*, 4–14.
- Guo, Q.; Zhang, Y.; Padash, A.; Xi, K.; Kovar, T. M.; Boyce, C. M. Dynamically Structured Bubbling in Vibrated Gas-Fluidized Granular Materials. *PNAS 2021*, 118 (35).

135b Novel Positron Imaging and Numerical Modelling of Bubbling Fluidised Beds: Toward an Improved Understanding of Pyrolysis-Based Waste Plastic Recycling

Dominik Werner, Kit Windows-Yule

Positron emission particle tracking (PEPT) is a technique which allows the three-dimensional dynamics of particulate and multiphase systems to be imaged with high temporal and spatial resolution. This high-resolution imaging, combined with its ability to penetrate optically-opaque materials, make PEPT highly valuable for the study of industry-relevant systems, where both the particulate media and the walls of the system are opaque. It is particularly well suited to applications such as gas-fluidised beds, whose rapid dynamics can be difficult to capture with more conventional methods such as x-ray CT.

In this talk, we focus on the application of PEPT to plastic recycling systems, in which fluidised-bed pyrolysis reactors are used to convert plastic waste into valuable products such as clean, low-sulphur fuels and virgin-quality plastic feedstocks. While this technology carries the potential to revolutionise our approach to the waste plastic crisis, the hydrodynamics of the systems used are highly complex and thus difficult to model, and thus optimise. As such, the transition from a promising proof-of-concept to an efficient, reliable, and thus commercially-viable system represents a significant challenge.

In this talk, we present key findings from a long-term project with industrial partners Recycling Technologies Ltd, providing an overview of the manner in which PEPT imaging has been combined with CFD-DEM and MP-PIC modelling to provide valuable, novel insight into the dynamics of their fluidised-bed-based plastic recycling systems. We will cover:

- How PEPT can be used to acquire valuable information regarding key hydrodynamic behaviours such as flow patterns and flow rates, bubble rise velocities, turnover rates, and both the rate and extent of mixing within a fluidised bed.
- The use of modular PEPT systems to perform *on site* imaging of active industrial pyrolysis reactors, extracting not only the above data, but also data pertaining to the rate of melting and volatilisation of plastics within the reactor.
- How these data can be used to calibrate and validate detailed CFD-DEM and MP-PIC simulations of the systems of interest.
- An overview of key results to date, including:
    - The influence of particle shape on the distribution of plastics through a binary sand-plastic bed.
    - The influence of plastic loading on flow dynamics and mixing rate.
    - The influence of distributor design on flow dynamics, mixing, and fluidisation quality.

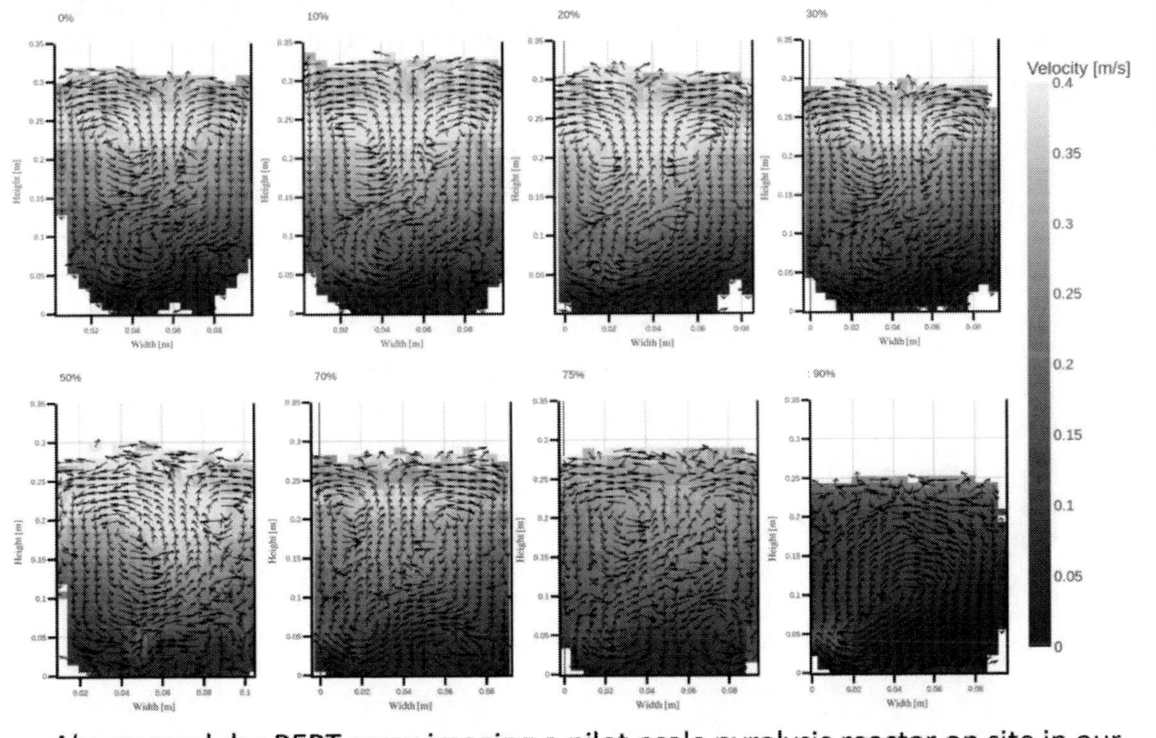

Above: modular PEPT array imaging a pilot-scale pyrolysis reactor on site in our industrial partner's HQ. *Below:* PEPT images illustrating influence of plastic loading on the flow patterns and flow velocities exhibited by a binary sand-plastic mixture under otherwise identical operating conditions.

135c Fluidized Bed Drying of Supported Catalysts: Effect of Process Parameters

Carlin Leung, Justin Adler, Tim A. G. Langrish, Benjamin Glasser

Drying of supported catalysts is an essential unit operation in the manufacturing of catalysts. Several different types of dryers are used for catalyst drying include belt dryers, rotary kiln dryers, and fluidized bed dryers. In fluidized bed drying, the heated gas flows through a bed of solids and fluidizes the bed, promoting high rates of heat and mass transfer between the gas and solids. Fluidized bed dryers allow for simultaneous spray impregnation and drying of supported catalysts, while promoting high moisture content uniformity through fluidization. Although several studies have been done on fluidized bed drying, general scale-up equations for fluidized bed drying of catalysts are lacking, as drying in a fluidized bed depends on the material properties and fluidization conditions. Thus, it is difficult to choose the operating conditions to optimize the drying process due to several phenomena occurring simultaneously, including heat and mass transfer, drying and fluidization. In general, one would like to have sufficient understanding of the process so that experiments done on a laboratory scale could be used to predict behavior at larger scales.

Experiments were carried out in a Glatt GPCG-1 fluidized bed dryer. Wetted catalyst supports were used as the experimental material. The rate of drying can be described by the change in moisture content of the bed over time. Experiments were done for different temperatures, gas flowrates and bed loadings. Model equations for predicting the moisture content of the bed were implemented using reference drying curves. We found that the model was generally capable of predicting the moisture content of the bed and the rate of drying for various process conditions. Moreover, for different fluidization conditions, we found that the model was capable of predicting the time to dry to 0% moisture content for cases with a delay in fluidization. We found that the model worked best when the majority of the drying took place in the unhindered drying period where the rate was roughly constant. However even when drying took place in the hindered drying period, the model was still able to make reasonably good predictions. We also found that there was better agreement between the model and experiments for high bed loadings where the number of transfer units was large. This work provides quantitative findings on how process conditions affect the rate of drying for a bed of catalyst supports with corresponding scale up equations.

135d Stirrer Design for Improving Fluidization of Cohesive Powder

Kaiqiao Wu, Rens Kamphorst, Jasper Ford, Gabrie M.H. Meesters, J Ruud Van Ommen

Fluidization is a popular technique to process solids for various applications in the industry, where outstanding interphase mixing and heat and mass transfer are required. Smooth fluidization is easily attained with sand-like particles, whereas it becomes rather troublesome to effectively aerate fine particles because of their intrinsic cohesiveness. Strong attractive forces dominate the collective behaviour of particles, resulting in the formation of agglomerates and gas channels which undermine enormously gas-solids interaction. Additional assistance is therefore necessary. A variety of methods, imposing different forces and energy sources to the system, have been demonstrated to effectively boost the fluidization of fine powders [1]. Among those, periodic stirring of fine powders has been shown to suppress channel formation, as well as to break up agglomerates [2]. Compared to other assistance methods, stirring is naturally in favour of scale-up practices concerning capital cost, operation and effective dimension. However, design of a stirrer is never a trivial task. Despite the popularity and advantages mentioned, only very few attempts are found on studying the stirred fluidization of cohesive powder. The impacts of stirrer designs remain largely unexplored.

In this work, we investigate the effects of stirring on the fluidization of cohesive powders with varying design parameters and operating conditions. The stirrers were customized and tested using a column of 5 cm inner diameter (Figure 1). A comprehensive comparison of fluidization behaviour is carried out across stirrer size, number, torsion angle, thickness and geometry. We evaluate the improvement based on the measurement of overall bed pressure drop, expansion and agglomerate sizes formed after fluidization. X-ray tomography technique was employed to reveal the system hydrodynamics and monitor channels for representative stirrer designs. Compared to the unassisted bed, stirring effectively improves the aeration of cohesive powder and mitigates channel formation. The data indicate volume-based stirrers surpass plane-based designs at assisting fluidization. Fluidization behaviour was observed nonlinearly associated with the design parameters, and the sensitivity analysis conducted forms a useful guide to design optimal stirrers for existing and future applications.

[1] J.R. van Ommen, J.M. Valverde, R. Pfeffer, Fluidization of nanopowders: a review, J. Nanopart. Res. 14 (2012) 737.

[2] S. Alavi, B. Caussat,. Experimental study on fluidization of micronic powders. Powder technol., 157(1-3) (2005) 114-120.

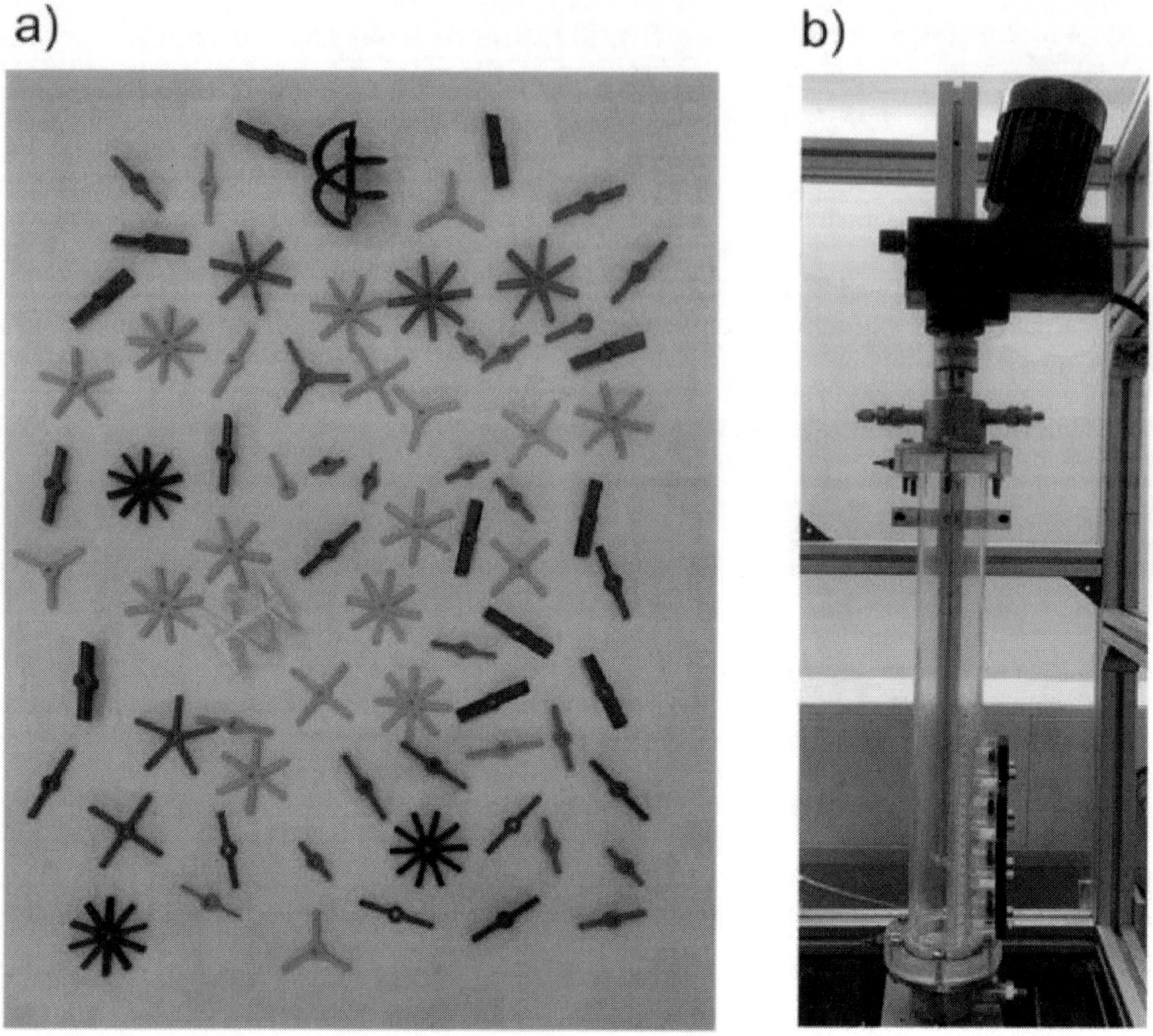

Figure 1: (a) Fused filament fabrication based 3D-printed stirrer with different properties, and (b) the experimental rig. Powder size is 7.5 μm and density is 1.9 g/cm^3.

135e A Numerical and Experimental Investigation of a High Temperature Fluidized Bed for Thermal Energy Transfer

Krutika Appaswamy, Jason Schirck, Zhiwen Ma, Aaron Morris

Several new thermal energy storage systems use silica sand as the storage medium and heat transfer fluid because sand is relatively inexpensive and stable at very high temperatures. In the energy system we investigate, surplus electricity from the energy grid is used to heat particles to 1200 °C. When required, the heated particles are fed to a pressurized fluidized bed heat exchanger, where the hot fluidization gases are then used to drive a turbine. A laboratory-scale pressurized fluidized bed (PFB) and discrete element model (DEM) are built to study the heat transfer and mixing phenomena in the PFB. NETL's open-source code, MFiX, is used as the computational tool to model the PFB. The MFiX-DEM is built to match the experimental setup with some reductions to improve the computational efficiency. The domain is a thin 3D rectangle with the height and length matching the experimental setup and a depth corresponding to 20 particle diameters. The PFB is run at 1100 °C and 10 bar pressure with the fluidizing air entering at 200 °C. The fluidization behavior, mixing, and particle heat transfer measurements are compared to the DEM model and existing empirical correlations. A good agreement of within 10% between the model and experimental is obtained for the bed pressure drop, defluidization curve, and particle heat transfer. In the experimental setup, we measure an initial temperature gradient within the bed. The temperature gradient arises due to higher temperature air on top of the bed relative to the bottom of the bed near the distributor. To approximately model the non-uniform initial bed temperature, the bed in the DEM simulation is initialized as having 3 sections of different temperatures. At superficial velocities very close to the minimum fluidization velocities, we observe that the bed is partially fluidized *i.e.*, the hot regions are fluidized and bubbling while the colder region remains unfluidized. This phenomenon is observed experimentally as well. It is also observed that cold particles tend to settle at the bottom of the bed, regardless of where they are at the beginning of the fluidization process. This has a significant effect on the bed mixing and the overall pressure drop in the bed and hence affects the thermal storage and heat transfer properties.

135f Analyzing Fluidized Bed Prediction Accuracy: From Modeling Framework to Material Properties

Casey LaMarche, Jia Wei Chew, Raymond Cocco

Achieving accurate predictions of fluidized bed behavior has been the focus of significant research efforts for the past several years. Understanding the sources of mismatch between predictions and experiments is clouded by a potential combination of various factors, e.g., missing physics, lack of relevant model validation, etc. In this work, fluidized bed predictions are compared to measurements for a wide experimental parameter space, several drag model formulations, and multiple numerical frameworks. In particular, prediction accuracy is assessed by comparing pressure gradient and entrainment rate measurements for bubbling and turbulent fluidized beds to predictions. Simulation predictions are compared against experimental measurements for Geldart Group A and Group B materials fluidized with multiple superficial gas velocities. In this investigation, the accuracy of multiple drag models are assessed and compared across the range of material properties and superficial gas velocity studied. Results from simulations conducted in the Eulerian-Eulerian (i.e., two-fluid model) and Eulerian-Lagrangian (e.g., MP-PIC) framework will be compared and discussed.

135g On the Behaviour of Highly Volatile Feedstocks in Fluidised Bed Reactors during Advanced Thermochemical Conversions

Stefano Iannello

For the last 200 years fossil fuels (e.g., coal, oil and natural gas) have provided us with cheap and convenient sources of energy. However, their massive utilization has also caused many problems to the environment, such as ozone depletion and global warming, associated with various emissions to land, water and air. Since the world energy consumption is destined to grow, finding effective, sustainable solutions to prevent the effects of anthropogenic global warming, will be the greatest challenge that our worldwide community face in the 21st century. Further to struggling with the effects of global warming, our society needs to face another great challenge: to develop economic and environmentally acceptable solutions for managing the ever-increasing volumes of municipal solid waste that are generated worldwide. Although recycling clearly has a critical role to play in reducing the amount of waste, there is further opportunity and environmental benefit in recovering energy from what might previously have been seen as materials destined for landfill. In particular, advanced thermochemical technologies (ATTs), like pyrolysis and gasification, have an important role to play in converting waste into clean energy or fuels, hence promoting the green energy transition. ATTs can be easily coupled with intermittent renewable energy, such as solar or wind, in order to enhance the performance of operation and further reduce the carbon footprint. In the last years, great interest has been showed towards biohydrogen as an alternative source of energy and ATTs in combination with renewable energies represent an attractive and sustainable alternative for the production of low-carbon hydrogen. All these aspects are of prime importance to carry forward the concept of the decarbonisation and meet the goal of 'net-zero' for 2050.

Fluidized bed reactors are among the most promising technologies for the thermochemical treatment of solid feedstocks for sustainable applications, due to their enhanced heat and mass transfer, and operating flexibility. Nevertheless, there are still unsolved challenges when operating with highly volatile solid feedstocks, such as biomass and waste. Because of their relatively low density, these materials experience axial segregation, tending to segregate along the bed height causing complications in the hydrodynamics of the bed. This behaviour is typically observed when operating with biomass. Moreover, volatiles released within the bed evolve in form of endogenous bubbles, which further enhance segregation of the feedstock. In addition, further issues arise because of the presence of synthetic polymers in waste, which tend to form agglomerates that sink to the bottom of the reactor. This phenomenon is detrimental as it causes de-fluidization of the bed and early shutdown of the operation. In both cases, solid and gas phases do not take advantage of the bed-to-fuel transfer phenomena, which are essential for high product yields and quality. All these aspects are essential part of the design of industrial fluidized bed units and relevant to all thermochemical conversions, since the evolution of volatiles is the first and common stage in pyrolysis, gasification and combustion operations.

The aim of this work is to provide a better understanding of the devolatilization behaviour and bed-feedstock interaction in fluidized bed reactors. This is done by investigating reacting biomass and plastic particles in a lab-scale fluidized bed reactor at high temperatures and different fluidization regimes relevant to industrial operations. The experiments were conducted in a fluidized bed operated at different temperatures ranging from 500 to 650 °C, which fall in the typical range of advanced thermochemical treatments conditions, and under either inert (pyrolysis) or semi-oxidizing (gasification/combustion) conditions. Beech wood and polypropylene spherical particles in the range of 8-12 mm were chosen to resemble the biomass and plastic components of waste, respectively. A novel non-invasive X-ray imaging technique, coupled with gas analysis measurements, has been used to investigate the thermal decomposition behaviour of a single particle after under-bed feeding via a purposely designed single particle injector. A small tracer of lead ranging from 1.5 to 2 mm was inserted into the half-drilled biomass and plastic particles in order to make them visible upon X-ray exposure. X-ray images have been collected with a time resolution of 36 frame-per-second and spatial resolution of about 1.6 mm/pixel. Algorithms of images post-processing and particle tracking have been developed in MATLAB environment in order to investigate the dynamic behaviour of the feedstock particle within the bed. A comprehensive assessment at minimum fluidization condition provided deeper insight about the devolatilization behaviour within the bed, which was found to be independent of the oxidizing nature of the fluidization medium. Size and void fraction distribution of endogenous bubbles were measured via X-ray imaging. Results showed that a fully developed endogenous bubble carries with it a volume of volatiles considerably larger that the volume predicted by a pseudo-first order rate law, which is usually used to describe the devolatilization behaviour of solid feedstocks. The evolution of void fraction of the endogenous bubbles released showed the presence of three main regions, namely cloud, wake and centre of the bubble, which mainly consists of gas produced form the devolatilization of the particle. In particular, the presence of the wake phase is known to be responsible of the upward movement of solids in fluidized beds. This finding confirms the existence of a lift effect acting on the feedstock, which can be assumed to occur when the bubble is fully formed and completely detached from the fuel particle. Gained knowledge provided by direct visualization from the collected X-ray images allowed a quantitative assessment of the lift effect acting on the reacting particle, which can then be used to develop a one-dimensional model to validate experimental results. Interestingly, the in-bed release of volatiles appeared to be independent of the presence of oxygen in the fluidizing medium. This finding was supported by visual evidence provided by X-ray visualization, which showed that endogenous bubbles size and frequency are independent of fluidizing gas properties. This behaviour was attributed to the controlling oxidation mechanisms of the volatile matter released, as demonstrated by the interchange phase theory of oxygen between emulsion phase and endogenous bubbles. The model showed accurate predictive capabilities for biomass particles at both minimum fluidization and bubbling regimes, while it failed in predicting the behaviour of plastic particles. The peculiar behaviour observed for this material was associated with its completely different mechanism of degradation, where the melting step was believed to have a great impact on both volatiles release and particle motion within the bed. The change in particle physical properties and formation of dense plastic-sand agglomerates might be responsible for the behaviour of this material.

Further experimental work will be conducted in order to gain a better understanding of the thermal decomposition behaviour of plastic materials in fluidized beds. The main focus will be on the mechanism of agglomeration of polypropylene particles at high temperatures. Experiments will be carried out by inserting a thermocouple inside the half-drilled sample and recording the temperature profile during the particle decomposition. This information will be essential to understand the thermal history of polypropylene under the thermal degradation conditions of fluidized beds, which details, unlike traditional thermogravimetric reactors, are still uncertain. Moreover, thermal information of the reacting particle might be crucial to understand the characteristic time of formation of plastic-sand agglomerates, and therefore explain their interaction with fluidized beds described previously. The collected data might be useful to build a mathematical model of plastic degradation in fluidized beds, in order to implement and improve existing numerical and CFD models.

Results obtained from this study highlighted the importance of developing systematic methodologies for the investigation of single particle devolatilization which, at present, appears to be the most effective approach to gain deeper insight on the interaction between feedstock and fluidized bed during thermochemical conversions. This is a central aspect for future development and exploitation of novel technologies, as well as improvement of existing ones, for thermal treatments of solid waste feedstock.

78a An Analog of Rayleigh-Bénard Convection in Vibrated Gas-Fluidized Beds

Qiang Guo, Yuxuan Zhang, Thomas M. Kovar, Kenan Xi, Chris Boyce

Fluidized beds have been subject to vibration to suppress bubbling, overcome cohesive forces and structure bubbling. Here, we demonstrate that vibrating fluidized beds with a low gas velocity can produce a fluidized state in which particles form convection cells of opposing chirality in which the size of the cells does not change as the system width is increased. We demonstrate that two-fluid modeling can reproduce this phenomenon and show that the convection is caused by a balance between buoyant and effective viscous forces, analogous to Rayleigh-Bénard convection in conventional liquids. We show vibration conditions can be manipulated to control convection patterns.

78b Fluid-like Phenomena in Binary Granular Materials: Revealing the Physics behind Rising Granular Bubbles and Splitting Granular Droplets

Jens Metzger, Ruben Strässle, Louis Girardin, Nicholas Conzelmann, Christopher McLaren, Christoph Müller

Agitated granular materials often exhibit similarities to fluids, such as breaking waves in granular shear layers[1], condensation-like droplets on tapped plates[2], and gas bubbles in gas-solid fluidized beds[3]. More recently, McLaren et al.[4] discovered a series of fluid-like phenomena that occur if a cluster of granular material is submersed in a granular material of different size and density and subjected to fluidizing gas and vertical vibration. If the cluster particles are larger and lighter than the surrounding bulk particles, a coherently rising structure forms ("granular bubble") reminiscent of a gas bubble rising in a liquid. Conversely, a cluster of particles smaller and denser than the bulk particles sinks and performs consecutive splits ("granular droplet") resembling a liquid drop sinking in a miscible fluid. Despite visual similarities to classical fluid phenomena, the emerging granular bubbles and droplets must be governed by distinct physics due to the ability of granular matter to solidify and the absence of surface tension between dry grains.

The present work elucidates the physics behind rising granular bubbles and sinking granular droplets and further explores the transition from rising to sinking. To this end, experiments in pseudo-2D vibro-gas-fluidized beds are combined with numerical simulations using computational fluid dynamics and the discrete element method (CFD-DEM). This combined approach allows insight into the interstitial gas flow and the network of inter-particle forces[5].

The results reveal a gas flow heterogeneity in the proximity of a granular cluster, where gas flows preferentially through regions of larger particles due to their increased permeability. This preferential gas flow is the driving force for the rising and sinking of the cluster since the heterogeneity locally affects the fluidization of particles and consequently their motion. Based on these results an analytical buoyancy model is derived to compile a dimensionless regime map that predicts the occurrence of granular bubbles and droplets, as well as additional regimes that have not been reported yet experimentally. By revealing the underlying physics of the motion of a granular cluster subjected to agitation, these findings have implications on the operation of industrial machinery such as vibro-fluidized beds, where binary or polydisperse granular materials tend to segregate.

1. Goldfarb DJ, Glasser BJ, Shinbrot T. Shear instabilities in granular flows. *Nature.* 2002;415(6869):302-305.
2. Duran J. Rayleigh-Taylor instabilities in thin films of tapped powder. *Phys Rev Lett.* 2001;87(25):254301.
3. Davidson JF, Harrison D, Carvalho JRFGD. On the liquidlike behavior of fluidized beds. *Annu Rev Fluid Mech.* 1977;9(1):55-86.
4. McLaren CP, Kovar TM, Penn A, Müller CR, Boyce CM. Gravitational instabilities in binary granular materials. *Proc Natl Acad Sci USA.* 2019;116(19):9263-9268.
5. Metzger JP, McLaren CP, Pinzello S, Conzelmann NA, Boyce CM, Müller CR. Sinking dynamics and splitting of a granular droplet. *Phys Rev Fluids.* 2022;7(1):014309.

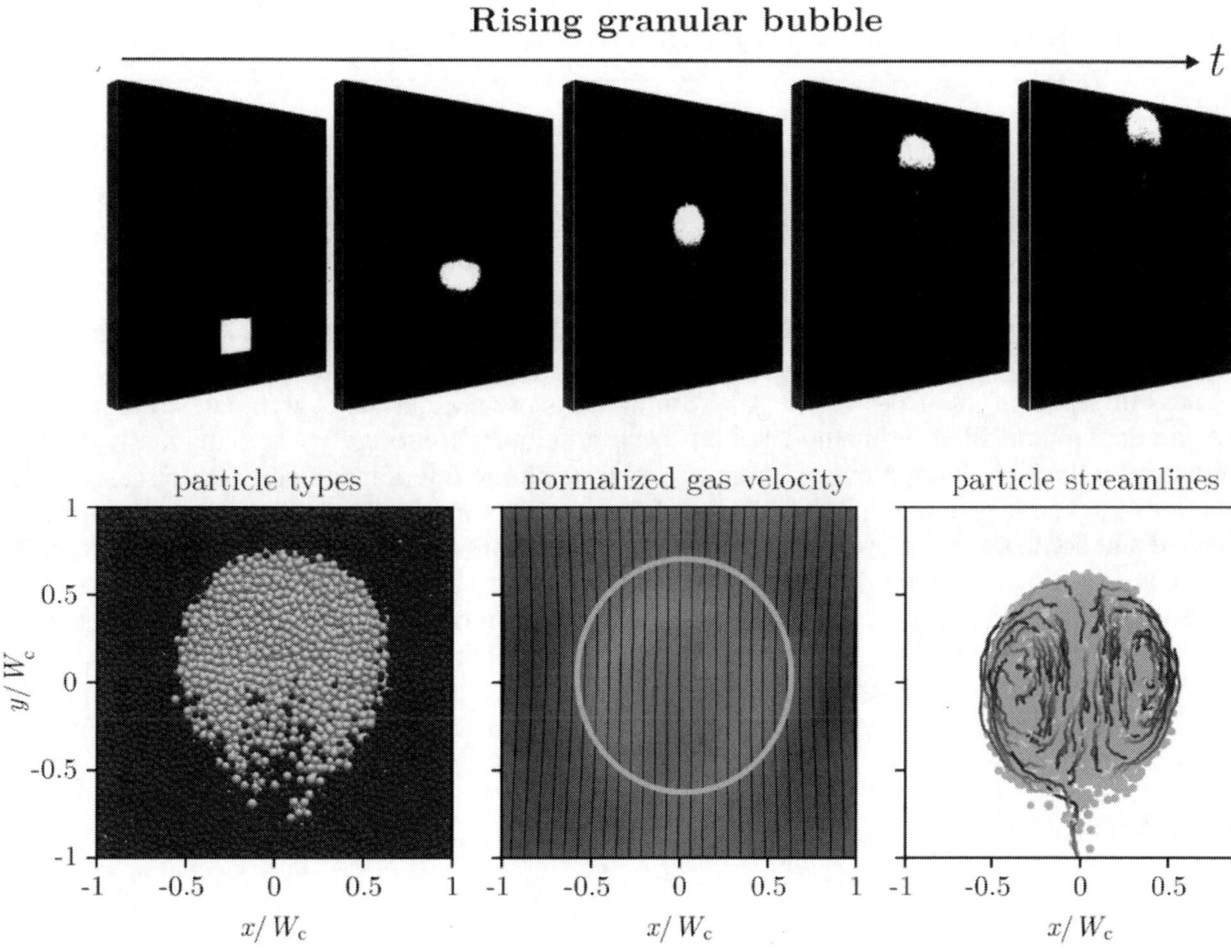
Rising granular bubble
t
particle types
normalized gas velocity
particle streamlines
1
0.5
0
-0.5
-1
y/W_c
-1
-0.5
0
0.5
1
x/W_c
x/W_c
x/W_c

78c Role of Mass Loading on the Onset of Flow Modulation in a Dilute Turbulent Slurry in Eulerian-Lagrangian Simulations

Mohamed Kasbaoui, Jonathan Van Doren

The modification of turbulent structures by point particles suspended in a fluid flow is of interest to pipeline design for oil and slurry transport. However, the point at which suspended particles start modulating the carrier flow remains poorly understood. To address this knowledge gap, we analyze the effect of point particles with different parameters on the carrier fluid statistics and near-wall coherent structures. The investigation is conducted in the configuration of particle-laden turbulent channel flow at a friction Reynolds number of 180, particle volume fraction varying from 1% to 6%, and mass fraction varying from 3% to 50%. Simulations use a Eulerian-Lagrangian strategy (EL) based on the volume-filtering method and has been previously thoroughly tested in massively parallel simulations of particle-laden flows. Depending on the volume fraction and particle diameter, the total number of tracked Lagrangian particles ranges between 180 thousand and 11 million. Statistics are collected and compared to statistics from auxiliary simulation with no particles. The simulations reveal that significant flow modification occurs only in cases where the mass loading approaches 50%. In such cases, we observe a reduction of the friction coefficient by about 15% and fluid mass flow rate by about 5%. Further, we show that particles with friction-based Stokes number below 1 remain randomly distributed throughout the channel, whereas particles with friction Stokes number about 7.5 display large inhomogeneity characterized by significant clustering and the tendency to accumulate near the walls.

This work is supported by ACS PRF# 62195-DNI9

78d Drag Reduction By Spherical Particles and the Role of Particle Clusters

Himanshu Dave, Mohamed Kasbaoui

We show that spherical particles dispersed in turbulent flow with high particle-fluid density ratio, yet low volume fraction, are able to reduce skin-friction drag by over 20%. The degree of drag reduction depends in large part on the solid mass fraction (mass loading) and particle Stokes number. This study is conducted using Euler-Lagrange simulations of four-way coupled turbulent particle-laden channel flow at friction Reynolds number 180, and particle-fluid density ration of 1000. To demonstrate the selective role of the Stokes number, two particle sizes are considered such that the friction Stokes number is 6 or 30. Additionally, the mass loading is varied from 0.2 to 1.0, corresponding to solid volume fraction between 2.4 10^{-4} and 1.2 10^{-3}. We show that Stokes 6 particles lead to drag enhancement for all mass loadings considered. The drag enhancement increases with Stokes number up to about 17% at mass loading 1.0. Conversely, Stokes 30 particles lead to drag reduction, which increases with increasing mass loading up to about 20% at mass loading 1.0. Our results show that particle clustering plays a significant role in the modulation of near-wall coherent structures and in fine drag reduction. At the low Stokes numbers, particles accumulate near the wall and clusters into small regions with no apparent connection to the coherent structures of the underlying carrier flow. At Stokes 30, the particles accumulate into very long ropes (exceeding 6000 wall units in length) along the walls of the channel. These ropes align with the near-wall low-speed streaks, suppress bursting, and contribute to the relaminarization of the near-wall flow. We show that in this drag reducing case, the spherical particles lead to similar effects as drag reducing polymers, namely, the the increase in the span-wise spacing of the near-wall low-speed streaks, and modification of the stress balance.

This work is supported by NSF CBET- 2028617 and ACS PRF# 62195-DNI9.

78e Maintaining Microstructure during Scaling in Circulating Fluidized Bed Risers

Ronald Breault

Over the years there have been numerous approaches to develop scaleup methodology for circulating fluidized beds. Many of these past approaches have relied on macroscopic definitions of similitude to define dimensionless groups such as Reynolds number, Froude number, density ratios, and dimensional ratios involving different length scales. These approaches have demonstrated various levels of success, but often fail when used outside the bounds of experimental datasets. The issue with these methods is that they rely on macroscopic similitude with the hopes that those conditions would provide microscopic similitude as it does in single phase systems where the Buckingham Pi techniques have given reasonable success.

Two of the main reasons for failure are 1) changing fluidization regimes and 2) changing material properties that result in the material falling into a different Geldart group. Both are certain to change the microstructure and changes to the microstructure affect the interphase transport properties of the reactor system, namely the mass transfer and heat transfer. This work presents a new approach towards scaling based upon maintaining similitude at the microscale. It reviews a novel dimensionless flow regime map developed from analysis of the deterministic chaos parameters and higher order moments to ensure microstructure similitude. It then presents an example of using the scaling method to obtain operating conditions for a scaled-up laboratory reactor. Four scaling cases from the literature are examined to understand if the microstructure was maintained.

78f Assisted Fluidization of Cohesive Powders: X-Ray Tomography Analysis

Kaiqiao Wu, Evert C. Wagner, Gabrie M.H. Meesters, J Ruud Van Ommen

Fine powders, with extremely high specific surface area, are of great interest for a wide range of applications in pharmaceutical, catalyst and functional materials production. Providing excellent homogenous mixing, heat and mass transfer and reduced pressure drops, fluidization is one of the most broadly used solid processing techniques nowadays in the industry. Nevertheless, fine powders tend to remain sub-fluidized when supplied with a steady gas flow, and constitute internal cracks and channels which bypass a significant portion of the fluidizing gas, resulting in poor interphase contacts. In addition, their cohesive nature promotes the agglomeration of individuals and, therefore, discourages overall mixing and solid circulation [1]. Over the last decades, several attempts were carried out to improve the fluidization of cohesive powders through additional manipulations on gas and solids. Typical actuators include mechanical vibration, gas pulsation, stirring and so forth [2]. Despite enhanced fluidization witnessed, the fundamental physics and hydrodynamics induced are yet far from being understood, which essentially complicates the relevant design, implementation, and scale-up practices.

X-ray computational tomography (XRT), as a powerful tool, offers unique insights into system hydrodynamics and direct visualization of local events [3]. In this work, we experimentally investigate the fluidization across different industrially relevant assistance, such as mechanical vibration and gas pulsation. Both XRT and pressure fluctuation measurements were conducted to monitor the temporal evolution of bed hydrodynamics and identify in-situ particle segregation, partial bubbling and channel formation (see Figure 1 for example). The observation is compared to the conventional fluidization, the degree-of-enhancement can be evaluated based on profiles of gas fraction, bed expansion and bed pressure drop. We also discuss the hydrodynamic events observed leading to the improved fluidization. These results help to optimize existing units of solid processing and develop an efficient process for catalyst production.

[1] J.R. van Ommen, J.M. Valverde, R. Pfeffer, Fluidization of nanopowders: a review, J. Nanopart. Res. 14 (2012) 737.

[2] V. Francia, K. Wu, M.-O. Coppens, Dynamically structured fluidization: Oscillating the gas flow and other opportunities to intensify gas-solid fluidized bed operation, Chem. Eng. Process. 159 (2021) 108143.

[3] J. Ma, J. R. van Ommen, D. Liu, R. F., Mudde, X. Chen, E. C. Wagner, C. Liang, Fluidization dynamics of cohesive Geldart B particles. Part I: X-ray tomography analysis, Chem. Eng. J., 359 (2019) 1024-1034.

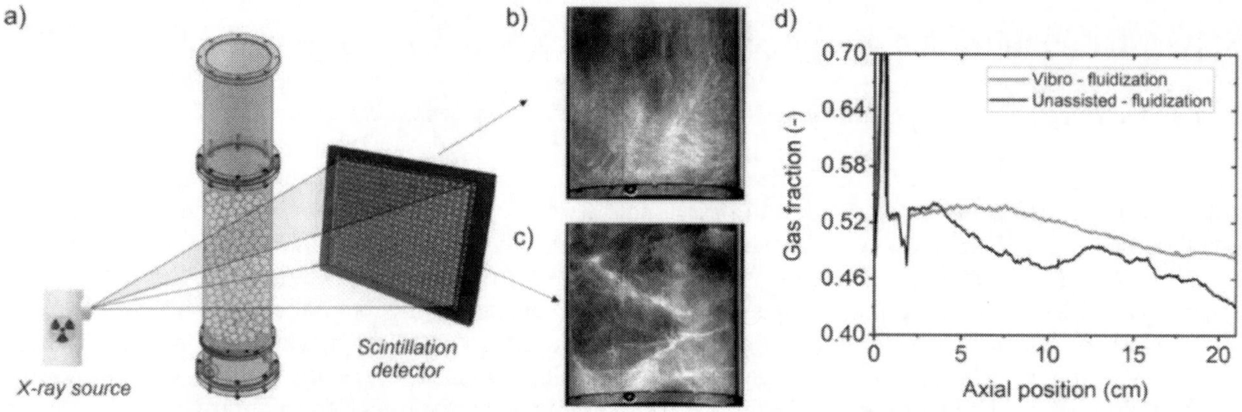

Figure 1 – (a) Schematic of X-ray tomography measurement setup, and examples of time-averaged profiles of gas channels created in the beds (b) under mechanical vibration and (c) without assistance; (d) corresponding the time-averaged axial gas fraction. Superficial gas velocity is 0.3 cm/s, powder size is 7.5 μm and density is 1.9 g/cm^3.

78g Experimental Investigation of Flow Behavior and Segregation for Nonspherical Particles in the Multi-Component Mixture through a Fluidized Bed Solids Mixer.

Sanjay Kumar Verma, Anshu Anand

This study aims to investigate the flow behavior of spherical and nonspherical particles in the gas-solid fluidized bed. Four different kinds of nonspherical particles such as oblate, prolate, cuboid, and hemisphere are considered to examine the fluidization behavior. These four nonspherical particles fluidized with the fine sphere in the gas-solid fluidized bed and recorded with a high-speed camera for post-processing. A high-speed camera is used to capture the images at 1000 fps. The particle image velocimetry (PIV) method is used to analyze the particle velocity using these images, which are generated from the high-speed camera. The consequences show that as increasing the superficial air velocity, the dynamic bed height of different mixtures varies. The velocity magnitude of the different mixtures is slightly varied but follows similar patterns. Increasing the aspect ratio of the nonspherical particle, both the segregation index and drag force reduce. Segregation index follows decreasing order of the nonspherical particles in the multicomponent mixture as oblate ($SI = 0.09$) > prolate ($SI = 0.07$) > cuboid ($SI = 0.06$) > hemisphere ($SI = 0.03$). It implies that as increasing the number of nonspherical components in the multi-component mixture, the mixing increases.

78h Strategies for Engineering Two-Dimensional Nanomaterials for Efficient Water Transportation

Edison Huixiang Ang

Engineering two-dimensional (2D) nanomaterials is important in membrane technology. The presentation will primarily focus on the strategies used to engineer 2D nanomaterials such as noble metals, MXenes, metal organic frameworks (MOFs), layered double hydroxide (LDH), and graphene for efficient membrane-based water purification. The correlation of the unique properties and rational design of 2D nanomaterials with respect to their performances in nanofiltration, solar still, and membrane distillation applications, will be described using experimental data and computational simulations. Finally, an innovative approach to using advanced three-dimensional (3D) printing technology (i.e. aerosol jet printing) of two-dimensional graphene-based materials for membrane technology will also be discussed.

Keywords: Nanotechnology; 2D nanomaterials; Engineering; Water; 3D printing

Acknowledgements

Prof. Edison Ang gratefully acknowledge the NAMIC-NTUitive, NIE-SUG, NIE AcRF and MOE Funds for supporting his research works.

References

1. M. Aizudin, R. Goei, A. J. Ong, Y. Z. Tan, S. K. Lua, R. P. Pottammel, H. Geng, X-L. Wu, A. I. Y. Tok, and **E. H. Ang***Journal of Materials Chemistry A, 10 (2022) 19612-19617.
2. **E. H. Ang**, Y. Z. Tan, J. W. Chew, Journal of Materials Chemistry A, 7 (2019) 10206-10211.
3. **E. H. Ang**, J. W. Chew, Chemistry of Materials, 31 (2019) 10002-10007.
4. **E. H. Ang**, L. Hong, Journal of Materials Chemistry A, 5 (2017) 20598-20602.
5. **E. H. Ang**, L. Hong, ACS Applied Materials andamp; Interfaces, 9 (2017) 28079-28088.

Graphical Abstract

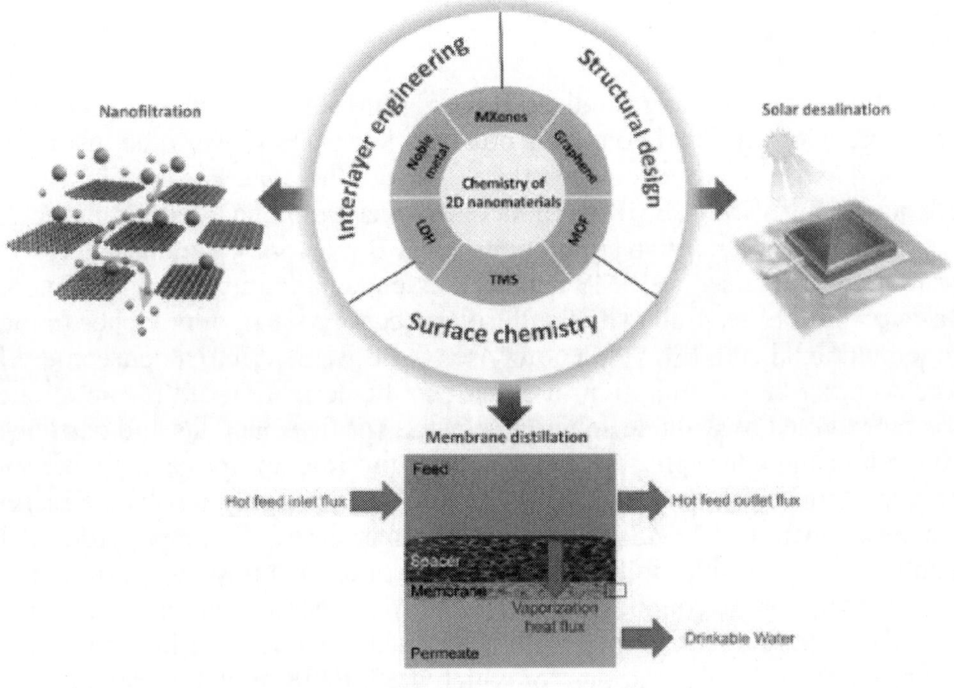

204a A Critical Assessment of the Energy Minimization Multi-Scale (EMMS) Model for Turbulent Fluidized Beds

Pedram Pakseresht, Jesse Capecelatro, Yuan Yao, Yi Fan, Jorg Theuerkauf, Lennard Lindmüller, Stefan Heinrich

Coarse grid simulations of circulating fluidized bed (CFB) reactors require an accurate estimation of the particle drag force. Conventional homogeneous drag laws are known to be insufficient for such configurations owing to the heterogeneous particle structure that emerges (e.g., clustering). The Energy Minimization Multi-Scale (EMMS) model has received much attention in recent years for capturing the effect of heterogeneity in simulations of CFB reactors. Originally derived for calculating the net particle force in CFB risers, EMMS has been extended and integrated within coarse grid simulations based on both Euler-Euler (EE) and Euler-Lagrange (EL) approaches to modify the drag force in each computational cell. EMMS decomposes the flow field into a dense and dilute phase that interact with each other. Many assumptions are made in the derivation of this model including (i) heterogeneity is represented by a single spherical cluster; (ii) invariant gas and particle superficial velocities for the dilute and dense phases; (iii) constant void fraction for each phase; and (iv) invariant particle drag force within each phase among others. Leveraging highly resolved EL simulations, we present a critical assessment of EMMS. The fundamental assumptions are probed in detail and model accuracy is quantified. We consider fully-developed homogeneous flow of solid particles settling under gravity. Strong interphase coupling results in the spontaneous generation of dense clusters that generate and sustain underlying turbulence. This setup is analogous to an individual computational cell in coarse-grained EE or EL simulations for which the EMMS model is typically applied. A structure tracking algorithm is applied to the highly-resolved EL data to measure cluster characteristics. Results show significant variation in the two-phase flow parameters, which differs from the key assumptions in the EMMS model. This work provides a framework for incorporating the observed variation and improving the standard EMMS model.

204b A New Model for Liquid Bridges in Wet Particles in Fluidisation

Ning Yang, Leina Hua, Raffaella Ocone

Addition of liquid into a fluidised bed of gas-solid flow is a common operation in industry. For instance, in fluidised catalytic cracking reactors the sprayed droplets introduced through atomisers may spread on the surface of hot catalyst particles and vaporise to enhance the cracking efficiency. The addition of liquid can give rise to strong cohesion between particles with the formation of liquid bridges that strongly affect the fluidisation.

In this paper we present a new model framework for liquid-bridge force is proposed to improve CFD-DEM simulation of wet particle fluidization. We integrated a temporal-scale criterion into the original spatial-scale threshold for liquid bridge force. Thus, a liquid bridge is allowed to form and remains stable when the normal relative velocity of two particles is lower than a critical value, in addition to the critical distance criterion. We show that the comparison between the predicted bubble behaviour and the measurements strongly indicates that there exists a critical particle velocity to impact lifetime of liquid bridge. If the relative velocity of particles is greater than the critical velocity of approach, the bridge might not experience a complete lifecycle. With the new model, the prediction of bubble properties including bubble centre, shape, and volume agrees well with the measurements in literature.

204c Comparison of Coarse-Grained CFD-DEM and KTGF Simulation for Industry Scale Fluidized Beds

Thomas Eppinger, Ravindra Aglave

Since its introduction [1], the Discrete Element Method (DEM) has proven to be a valuable method for the analyzing and understanding particulate flows. Supported by the continuously increasing computational power, CFD-DEM simulations have found their way into the chemical and process industry for various applications like solid suspension in mixing vessels, fluidized and spouted beds, granular transport and coating applications in rotary drums [2,3]. The major shortcoming of DEM, however, is its computational cost that increases with the amount of particles involved, their material properties (stiffness) and size. This hinders the application of CFD-DEM simulation to large-scale systems of industrial size. To overcome this shortcoming a coarse grain (CG) model has been described [4]. Using straightforward scaling rules, a group of particles gets replaced by a representative coarse parcel. This effectively reduces the number of particles that need to be processed and subsequently shortens the computational time. On the other hand, they introduce a modeling error into the simulation. In a previous study we were investigating different coarse graining models and quantify the effect on different parameters and their fluctuation like pressure drop and expansion height in a fluidized and spouted bed and we were comparing the simulation with experimental results. In general, it can be stated that bed expansion and pressure drop are well predicted for all investigated CG methods but depending on the CG factor the dynamic of the system cannot be captured. [5]

An alternative approach is the kinetic theory of granular flow based on the eulerian multiphase model. The different phases are treated as interpenetrating continua, but additionally the intraphase interaction of the dispersed solid phase is considered. A validation of different KTGF-models and the influencing model parameters was recently presented. [5]

In this work we are comparing coarse-grained CFD-DEM, EMP-KTGF and experimental results for an industrial scale fluidized bed in terms of averaged as well as dynamic data. Results of different modeling aspects are quantified and discussed with a focus on model modifications to capture the mesoscale effects. Finally, the effect on runtime and ways to reduce it ate discussed.

References:

1. Cundall, O.D.L. Strack, Geotechnique, 29 (1979), 47
2. Eppinger, O. Baran, R. Aglave, S. Lo, Simulating Solid Suspension in Stirred Vessels with a Fully Coupled CFD-DEM Algorithm (161d), AIChE Annual Meeting 2017, Minneapolis
3. Baran, R. Aglave, M. Tandon, A. Karnik, S. Lo, Numerical Simulation of Dense Gas-Solid Fluidized Beds: Comparison Between Eulerian Multiphase and Discrete Element Methods, AIChE Annual Meeting 2015, Salt Lake City, UT.
4. Sakai, Y. Yamada, Y. Shigeto, K. Shibata, V.M. Kawasaki, S. Koshizuka, Numerical Methods in Fluids 64 (2010), 1319
5. Eppinger, T., Tourani, C., Aglave, R., Quantifying Effects of Coarse-Grain Model Parameter in CFD-DEM Simulations of Fluidized and Spouted Beds and Comparison Against Eulerian Multiphase Simulation Results, AIChE Spring Meeting 2019, New Orleans

204d Concentric-Tube Internal Loop Airlift Reactors for Microalgae Cultivation: A Review

Lifeng Li, Xiaoyun Xu, Wujun Wang, Raymond Lau, Chi-Hwa Wang

Microalgae are widely used in applications such as aquaculture, environment engineering, food, and pharmaceuticals [1]. Recently, microalgae technologies attract a surging attention due to the great capabilities of producing bio-fuels and sequestrating carbon dioxide, thus meeting the increasing demand of sustainable aviation and marine fuels production and carbon reduction [2]. An enclosed, controlled, nourished, and illuminated culture environment for the production of biomass can be provided by a photobioreactor, particularly the concentric-tube internal loop airlift reactor. This type of reactor offers the advantages of simple construction, easy operation, low power consumption, defined fluid flow patterns, and creation of light/dark cycles.

The concentric-tube internal loop airlift reactors are common reactor configurations adopted for microalgae cultivation, as well as various chemical and biochemical processes. Despite the fact that this type of reactor has been proposed for a few decades and many research studies were conducted, there is no a research study which provides a thorough introduction of the basic concepts, configurations, and numerical and experimental methodologies for investigating the reactor performance. Hence, the present work provides a comprehensive review of the reactor basic concepts, classifications, theoretical background, performance metrics, numerical and experimental methodologies, and parametric analyses on the effects of reactor design parameters, as well as recent advancement techniques.

References:

[1] P. Spolaore, C. Joannis-Cassan, E. Duran, A. Isambert, Commercial applications of microalgae, Journal of Bioscience and Bioengineering 101 (2) (2006) 87–96.

[2] G. Satpati, R. Pal, Microalgae-biomass to biodiesel: a review, Journal of Algal Biomass Utilization 9 (4) (2018) 11–37.

204e Model Parameters Calibration for DEM By Macroscopic Characterization of Granular Material

Filip Francqui, Salvatore Pillitteri, Aurelien Neveu, Geoffroy Lumay

DEM simulations are powerful tools to model the powder behaviours. Especially, they allow to simulate powder at large scale and to reproduce granular materials behaviour in an industrial process. Reproducing a complete industrial process in laboratory can be complex and expansive. When this process is modeled by an experimental setup, that reproduces the industrial device at a lower scale, correlating the results with what is observed at larger scale can be complex or irrelevant. The advantage of the simulations is that they can model these complex processes at a similar scale. However, it can be hard to know the adequate microscopic parameters to fix in the simulation (coefficients of friction, coefficients of restitution, cohesive interactions, ...) to reproduce what is observed experimentally.

Some measurements can be performed at the scale of the grains, like AFM techniques, to obtain these parameters, but these methods can be time consuming and expansive. We propose a top-bottom method based on the characterization workflow of a real powder in a rotating drum. This geometry is adequate to characterize the cohesiveness and the flowability of a powder and can be easily simulated. Therefore, experimental results can be directly compared with the simulations to make a validation of the selected parameters. We present case studies obtained with the GranuDrum, an instrument that reproduce the rotating drum geometry. Moreover, we discuss the perspectives that this kind of methodology can bring for a better understanding of powder flowability and process improvement.

204f Heat and Mass Transfer within Dynamically Structured Bubbling Fluidized Beds Subject to Vibration: A Two-Fluid Modeling Study

Qiang Guo, Shawn Chiu, Wei Da, Chris Boyce

Gas-solids bubbling fluidized beds are widely used in a range of chemical and petroleum processes. The dynamic coalescence, split up, and collapse of bubbles in fluidized beds are mathematically chaotic and hard to predict, which create significant difficulties in scale-up from laboratory- to pilot- to industrial-scale. In the past few decades, many different ways have been proposed to manipulate the overall bubble dynamics in fluidized beds to introduce more structures that can simplify the overall scale-up process. Among all of the methods used, it has been found that using pulsed gas flow or vertical vibration can cause otherwise chaotic gas bubbles to form a structured triangular pattern, which repeats itself every two gas oscillation or vibration periods. Therefore, the induced bubbling structure is highly predictable. Further, the characteristic bubble size and bubble distance remain unchanged when the system width is changed, and using a layered configuration with multiple stages of gas distributors can maintain the ordered bubbling structure to a higher particle height. The possibility to maintain the structured bubbling pattern with the system width and system height changed opens great opportunities to address key issues in scaling up traditional gas-solids bubbling fluidized beds. Based on Computational Fluid Dynamics-Discrete Element Method (CFD-DEM) simulations and advanced measurement techniques, structured bubbling and its effect on the particle mixing and gas-solids contact patterns have been well characterized. However, the effect of the structured bubbling on heat and mass transport remains unclear. Herein, using two-fluid modeling (TFM), we studied the heat and mass transport within dynamically structured bubbling fluidized bed and compared with free bubbling fluidized bed. The heat transfer was initiated by cooling particles at high thermodynamic temperature using cool inlet gas flow and the mass transfer was conducted by removing the coke deposited on the particles with air combustion. The results showed that compared with free bubbling fluidized bed, the particles in the dynamically structured bubbling fluidized bed show a more uniform temperature distribution and more uniform coke concentration, suggesting that the dynamically structured bubbling fluidized bed has the potential to be used in in a range of applications, in which a narrow temperature distribution or a narrow species distribution within the particle phase is a key consideration.

79a Thermomechanical Properties of Polyurea Nanocomposites over Extreme Strain Rates

Jessica Kopatz, Elizabeth Jones, Brett Sanborn, Justin Wagner, Christopher Riley, Christine Cardinal Roberts

Polyurea is commonly used as an encapsulant, a coating, and/or a binder to increase durability, corrosion, or impact resistance of surfaces. Unlike materials with similar chemistries, such as polyurethane, polyurea has a phase-separated microstructure consisting of hard-segments that are dispersed throughout a soft-segment matrix. This phase separation and hydrogen bonding in the material results in exceptional impact resistance and toughness, even at high strain rates and high strain levels. Here, we demonstrate enhanced thermal and mechanical properties of polyurea through the incorporation of nanoparticles. Surface modification of the nanoparticles was optimized to promote homogeneous particle dispersion, maximum particle volume fraction, and particle-matrix adhesion. Particle-matrix compatibility significantly affects the ultimate tensile stress for the nanocomposites. The dispersibility of particles is measured via rheology while the phase segregated microstructure and particle segregation is observed via atomic force microscopy. The mechanical performance of polyurea nanocomposites is compared to that of unfilled polyurea over an extreme range of strain rates (0.1-10,000 Hz) via dynamic mechanical analysis, Hopkinson bar, and shock tube testing, demonstrating the ability of polyurea nanocomposites to damp vibrational energies over a wide range of rates. Finally, highest strain rate insults were imposed in the Sandia National Laboratories SPHINX electron beam and Z-machine X-ray radiation pulsed power facilities, where the damping of mechanical energies of polyurea was measured to strain rates of 10^6 s^{-1} by imposing a thermomechanical shock in the material. Assessment of the damage to the polyurea as a result of electron beam radiation shows an exceptional resistance of the nanocomposites to radiation and thermal damage.

SNL is managed and operated by NTESS under DOE NNSA contract DE-NA0003525

79b Light Extinction By Agglomerates of Gold Nanoparticles: A Plasmon Ruler for Sub-10 Nm Interparticle Distances

Georgios Kelesidis, Daniel Gao, Fabian Starsich, Sotiris E. Pratsinis

Plasmon rulers relate the shift of resonance wavelength, λ_l, of gold agglomerates to the average distance, s, between their constituent nanoparticles. These rulers are essential for monitoring the dynamics of biomolecules (e.g., proteins and DNA) by determining their small (<10 nm) coating thickness. However, existing rulers for dimers and chains estimate coating thicknesses smaller than 10 nm with rather large errors (more than 200%). Here, the light extinction of dimers, 7- and 15-mers of gold nanoparticles with diameter, d_p = 20 - 80 nm and s = 1 - 50 nm is simulated by interfacing discrete element modeling (Kelesidis et al, 2017) with discrete dipole approximation (Kelesidis and Pratsinis, 2019). The evolution of gold light extinction during agglomeration reveals that λ_l increases up to 680 nm as single gold nanoparticles coagulate to 15-mers, in excellent agreement with data from UV-visible spectroscopy. The λ_l shift of gold dimers, 7- and 15-mers increases with decreasing s and can be described by a universal power law resulting in a new plasmon ruler that enables the estimation of gold nanoagglomerate s (Fig. 1), in excellent agreement with microscopy data. So, the new plasmon ruler obtained here for agglomerates of gold nanoparticles can be used instead of tedious microscopy measurements to determine the thickness of sub-10 nm organic and inorganic coatings, facilitating the detection of biomolecules, such as proteins and DNA (Kelesidis et al., 2022).

References:

Kelesidis, G. A., Goudeli, E., and Pratsinis, S. E. 2017. Morphology and mobility diameter of carbonaceous aerosols during agglomeration and surface growth. Carbon 121:527-535, doi.org/10.1016/j.carbon.2017.06.004.

Kelesidis, G. A., and Pratsinis, S. E. 2019. Soot light absorption and refractive index during agglomeration and surface growth. Proc. Combust. Inst. 37: 1177-1184, doi.org/10.1016/j.proci.2018.08.025.

Kelesidis, G. A., Gao, D., Starsich, F.H.L., and Pratsinis, S. E. 2022. Light Extinction by Agglomerates of Gold Nanoparticles: A Plasmon Ruler for Sub-10 nm Interparticle Distances. Anal. Chem. 94:5310-5316, doi.org/10.1021/acs.analchem.1c05145.

Figure 1. Au agglomerate λ_l as a function of normalized interparticle separation, s/d_p, estimated by DDA using dimers of monodisperse (squares and inverse triangles) and bidisperse (circles) nanoparticles and DEM-derived 7- (diamonds) and 15-mers (triangles) of monodisperse NPs with d_p = 20–50 (squares, circles, diamonds, and triangles) or 75 and 80 nm (inverse triangles) and s = 1–50 nm. A new plasmon ruler (solid line and shaded area) is derived by regressing the DDA-derived λ_l evolution.

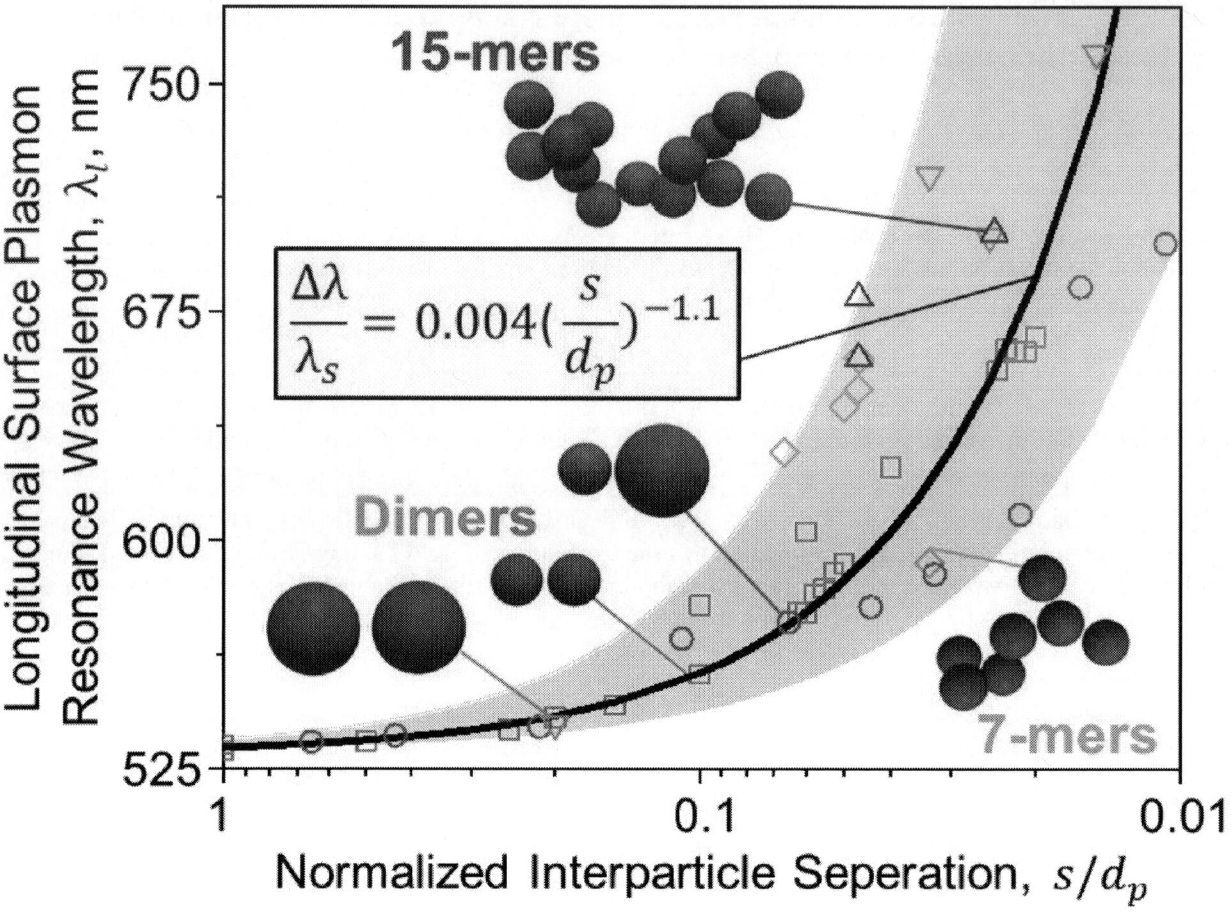
15-mers
Dimers
7-mers
$\frac{\Delta\lambda}{\lambda_s} = 0.004(\frac{s}{d_p})^{-1.1}$
Longitudinal Surface Plasmon Resonance Wavelength, λ_l, nm
750
675
600
525
1
0.1
0.01
Normalized Interparticle Seperation, s/d_p

79c Photothermal Conversion Efficiency of Multi-Color Emissive Carbon Dots: A Chemical and Thermal Analysis

Salar Balou, Aashish Priye

A series of highly stable, cost-effective, and environmental-friendly multi-color emissive carbon dots with systematically tailored chemical and physical properties were synthesized through a modified acid reagent engineered hydrothermal synthesis protocol. Detailed optical and thermal characterizations together with computational modeling revealed their remarkable photothermal (light-to-heat) conversion performance and the underlying mechanisms governing this observation were studied and discussed. Results showed a significantly higher light-to-heat conversion in red color emitting carbon dots compared to their blue, cyan, green, and yellow counterparts. Apparently, more acid treatment during the fabrication of carbon dots enhances the aromatization and introduces stronger electron donating functional groups (such as C-O-C, and C=O groups) into the lattice which gives rise to a more conjugated and oxidated carbon dot structure. In return, the HUMO and LUMO band gaps will get smaller which destabilizes the electrons in π-π^* orbitals and significantly increases the frequency of electron-phonon coupling in carbon dots. This phenomenon (similar to localized surface plasmon resonance) initiates a thermal vibration in the carbon dot lattice (the phonon-phonon coupling) which eventually dissipates heat into the surrounding media that we interpret as temperature increase in macroscopic scale. The highest thermal gradient observed was 11 °C/s in red color emitting carbon dots when illuminated with a blue laser (350 mW, continuous beam). Finally, these nanocarbons were employed as efficient photothermal nanomaterials to achieve ultrafast thermocycling for quantitative real-time PCR owing to their potential of implementing 30 cycles of 95 °C to 60 °C in less than 10 minutes. This work will set the first stage for bandgap engineering to utilization of carbon dots as efficient and effective nanomaterials for forward-looking photothermal applications and studies.

79d *Probing Nanoparticle Interaction with Artificial Cell Membranes Via a High-Throughput Fluorescent Liposomal Leakage Assay*

Alison Lui, Kranthi K. Mandadapu, Markita Landry

Nanomedicines and nanotherapeutics consist of loading biomolecular cargo in or on a nanocarrier, followed by co-incubation or injection into a system of interest, and subsequent integration or uptake past the cell membrane and into a cell. On the scale of the cell membrane, few studies have demonstrated the degree to which qualities of a nanoparticle such as size, functionalization, and coronal surface layer composition fundamentally change the uptake and interaction with the bilayer itself. In this study, we have demonstrated the capability of a high-throughput fluorescence assay to measure how physico-chemical qualities of nanoparticles enable or spurn bilayer interaction. We demonstrate the breadth of this assay by considering a wide variety of nanoparticle compositions and functionalizations. Further, we make preliminary strides to explore the effect of an adsorbed protein corona and thus elucidate the importance of the cellular environment in impacting membrane-nanoparticle interactions.

This fluorescence self-quenching assay relies on (5)6-carboxyfluorescein (CF), a membrane impermeable dye loaded into large unilamellar vesicles (LUV) at a concentration around 40 mM where these tightly packed dye molecules show minimal fluorescence. Release of the CF (as a result of nanoparticle intercalation and disruption at the LUV bilayer) results in an immediate strong fluorescent signal. Comparing the induced dye release over time upon interaction with therapeutically relevant functionalized nanomaterials informs understanding of their compatibility with phospholipid bilayers and exosome-like structures.

A wide range of nanoparticles were considered including (i) 10 nm diameter gold nanospheres functionalized with citrate or an oligonucleotide sequence and (ii) colloidally stabilized single walled carbon nanotubes. We discuss approaches to controls and data processing to account for some inherent nanoparticle light absorption in this fluorescent reporter system. We also incubate select nanoparticles with relevant biological proteins or serum to form an adsorbed protein corona. We then apply our assay to begin elucidation of the degree to which complex biological environments mediate interaction.

Translation of our inferences in nanoparticle-bilayer interaction kinetics to nanomedicine can inform optimal design of targeted delivery agents.

136a Toxicity of Orally Administered Magnetic Imaging Nanoprobes in an Inflammatory Intestinal Epithelial Cell Model

Shno Asad, Christel A.S. Bergström, Alexandra Teleki

Introduction: SiO_2-coated superparamagnetic iron oxide nanoparticles (SPION) can be efficiently produced in a scalable, single-step flame process.[1,2] The SiO_2-layer on the particle surface facilitates subsequent chemical modifications, and the magnetic iron oxide core makes the SPIONs suitable as MRI contrast agents. We hypothesized that functionalizing SPIONs with ligands that target overexpressed biomarkers in the gastrointestinal tract (GIT) can allow non-invasive imaging of diseases such as inflammatory bowel disease (IBD) *in vivo*. Biomarkers in preclinical IBD models include ICAM1,[3] which could be targeted using functionalized SPIONs. However, exposing damaged and inflamed cells to such functionalized nanoparticles might have the potential to cause immunotoxicity and compromise their safe use as contrast agents.

Methods: Pure iron oxide and SiO_2-coated iron oxide nanoparticles are made through flame spray pyrolysis.[1] Monoclonal antibody anti-ICAM1 is conjugated onto the SiO_2-coated particle surface using click chemistry to target the IBD associated biomarker ICAM1. Filter-grown Caco-2 cells are treated with a mixture of inflammatory inducing agents to resemble the inflamed state of IBD. Both the antibody-functionalized, non-functionalized and pure SPION particles are evaluated for their cytotoxic and genotoxic effects, as well as the morphological changes in the cell monolayer. Cytotoxicity and genotoxicity is examined using CellTiterGlo assay and Comet assay, respectively. Cell integrity is examined by measuring permeability of Lucifer yellow across the cell monolayer.

Results: Global proteomics of inflamed Caco-2 cell model has shown the upregulation of biomarkers, such as ICAM1, in the apical side of the plasma membrane.[3] Caco-2 cell viability was not altered by the exposure to SiO_2-coated SPIONs, which indicates an increased biocompatibility due to the coating layer. As expected, the inflamed cell model demonstrated a higher leakiness across the monolayer compared to the healthy cells. However, the monolayer integrity of the inflamed cell model was not found to be significantly worsened after 24 h of exposure to the SiO_2-coated SPIONs. These results suggests that SPIONs used as MRI imaging probes would not impair the inflamed state during an active inflammation.

Conclusions: Adverse toxic effects of nanomaterial based imaging probes for use *in vivo* is important to assess. Evaluation of cytotoxic and genotoxic effects, as well as changes in permeability of inflamed Caco-2 cell monolayers provide an estimation of the safety to use functionalized SPION as a diagnostic tool *in vivo*.

References: [1] Teleki, A., M. Suter, P.R. Kidambi, O. Ergeneman, F. Krumeich, B.J. Nelson, and S.E. Pratsinis, *Chem. Mater.* **21**, 10, (2009). [2] Teleki, A., F.L. Haufe, A.M. Hirt, S.E. Pratsinis, and G.A. Sotiriou, *RSC Adv.* **6**, 26, (2016). [3] Asad, S., Wegler, C., D. Ahl, C.A.S. Bergström, M. Phillipson, P. Artursson and A. Teleki, *J. Pharm. Sci.* **110**, 1 (2021).

136b Making Nanostructured Materials for Health Applications Using Atomic Layer Deposition

J Ruud Van Ommen, Volkert van Steijn, Antonia G. Denkova, Alina Rwei

While nanotechnology could potentially play an enormous role in the medical field, it is often not straightforward to make nanostructured materials for pharma and medical devices in a scalable manner. We will show that variations of atomic layer deposition (ALD) – a coating technique that has really taken off in the semiconductor sector – can be used to make nanostructured materials for several health applications in a scalable way.

Modifying the surface of grains of powder – as used for example in dry powder inhalers for treating respiratory diseases – can increase the powder flowability while also adding a controlled release functionality. In this way, the powder is used more effectively and side-effects in the patient can strongly be reduced [1,2].

In targeted radionuclide therapy a major hurdle is the dependence on a very limited number of nuclear reactors worldwide to produce these radioisotopes. ALD can provide a way to prepare radionuclide generators, that can be placed in hospitals, providing on-site and on-demand supply. By modifying sorbent particles for isotope separation by ALD, we have made the first step towards ^{99}Mo/^{99m}Tc radionuclide generators [3]. And by Lu ALD on nanoparticles, we have taken a step in producing ^{177}Lu radionuclides [4].

We also apply ALD to other substrates than powders. Polydimethylsiloxane (PDMS) is a widely used material for lab-on-a-chip and organ-on-a-chip devices. It is very versatile, but the drawback is its low stability, especially when in contact with organic solvents. We will show that modifying the surface with ceramic oxide using ALD is a way to get much more durable microfluidic devices, usable for a range of health applications.

[1] Zhang, D., La Zara, D., Quayle, M. J., Petersson, G., Van Ommen, J. R., andamp; Folestad, S. (2019). Nanoengineering of crystal and amorphous surfaces of pharmaceutical particles for biomedical applications. *ACS Applied Bio Materials*, *2*(4), 1518-1530.

[2] La Zara, D., Sun, F., Zhang, F., Franek, F., Balogh Sivars, K., Horndahl, J., Bates, S., Brännström, M., Ewing, P., Quayle, M. J., Petersson, G., Folestad, S. andamp; van Ommen, J. R. (2021). Controlled Pulmonary Delivery of Carrier-Free Budesonide Dry Powder by Atomic Layer Deposition. *ACS Nano*, *15*(4), 6684-6698.

[3] Moret, J. L. T. M., Alkemade, J., Upcraft, T. M., Oehlke, E., Wolterbeek, H. T., van Ommen, J. R., andamp; Denkova, A. G. (2020). The application of atomic layer deposition in the production of sorbents for 99Mo/99mTc generator. *Applied Radiation and Isotopes*, *164*, 109266.

[4] Moret, J. L., Griffiths, M. B., Frijns, J. E., Terpstra, B. E., Wolterbeek, H. T., Barry, S. T., Denkova, A.G. andamp; Van Ommen, J. R. (2020). Lutetium coating of nanoparticles by atomic layer deposition. *Journal of Vacuum Science andamp; Technology A: Vacuum, Surfaces, and Films*, *38*(2), 022414.

136c Rational Design and Interspecies Translation of Glucose-Responsive Insulins and Insulin Delivery Systems By Computational Modeling

Sungyun Yang, Jing Fan Yang, Xun Gong, Michael Strano

Glucose-responsive insulin (GRI) is a type of 'smart' insulin therapy for diabetic patients to mitigate the risk of the tight therapeutic range of blood glucose concentration in typical insulin replacement regimens. The roles of GRI are to prevent the hypoglycemic risk of diabetic patients from exogenous insulin treatment and improve patients' lives by avoiding regular subcutaneous insulin injections, which are major potential sources of user error. Researchers have investigated various novel approaches to realize GRIs for decades. These approaches can be classified into three categories: intrinsic GRIs, glucose-responsive particles, and glucose-responsive insulin delivery devices. Intrinsic GRIs are modified insulin analogues that exhibit inherent glucose-responsiveness. Glucose-responsive particles are freely circulating micro- or nanoparticles that release insulin during hyperglycemia. Finally, glucose-responsive insulin delivery devices are transdermal insulin delivery devices only activated during hyperglycemia. We developed a computational model named PAMERAH (Pharmacokinetic Algorithm Mapping GRI Efficacies in Rodents and Humans), which enables the assessment of GRI therapeutic efficacy and translatability between humans, rats, and mice. These models provide an important tool for the evaluation of clinical trial outcomes. PAMERAH is constructed of the compartmental physiological model of humans and rodents, with the additional GRI compartments are embedded. Our previous studies focused on intrinsic GRIs – here, we expand our model to other types of glucose-responsive systems: glucose-responsive particles and devices. The model examines efficacies after subcutaneous injection of GRIs over 24 hours, and we map the therapeutic range of GRIs based on dose and kinetic parameters. With the PAMERAH construct, we demonstrate that different classes of GRI exhibit different translatability patterns between humans and rodents, and these results show the capability of our model to simulate complicated therapeutic constructs. Hence, we envision PAMERAH as an effective tool for GRI development and a route toward realizing model guided clinical treatment for humans.

136d Processing of Lipid Nanoparticles for RNA Encapsulation

Navid Bizmark, Satya Nayagam, Dawei Zhang, Sujit Datta, Rodney Priestley, Robert K. Prud'homme

Lipid nanoparticles (LNPs) have recently emerged as effective vehicles for ribonucleic acid (RNA) delivery in vaccine formulations. LNP-based RNA vaccines for COVID-19 are among the most recent and critically important examples of such advanced formulations. While there has been extensive research on lipid structure, lipid formulations, and RNA modifications to optimize performance, there has been less work on how processing affects LNP structure and properties. In this work, we have used a Confined Impinging Jet and a Multi-Inlet Vortex Mixer to produce LNPs using Flash NanoPrecipitation (FNP). In these turbulent mixers (Reynolds number>5000), an organic solvent stream—containing the lipids—and an aqueous anti-solvent stream which may contain RNA are rapidly mixed. Through this scalable process LNPs are formed only at the proper compositions and flow rates of the fluid streams. We have explored the effects of buffer pH, ionic strength, and lipid concentrations on the key properties of LNPs, including size, zeta potential, and colloidal stability. We show how these parameters can be tuned to produce 60 to 80 nm LNPs—an optimal size for the best *in vivo* practices. Moreover, we have maximized the loading of RNA in LNPs under controlled electrostatic and steric interactions. These findings are critical to produce RNA-loaded LNPs, and further develop therapeutic formulations.

136e Tailoring Spions through Polymer Coating Modifications for Cell Labeling and Tracking with Magnetic Particle Imaging (MPI)

Hayden Good, Rita Das Mahapatra, Sitong Liu, Carlos Rinaldi-Ramos

The use of adoptive T cell therapy has growing interest due to its promise for cancer immunotherapy. Studies suggest treatment outcomes correlate with the localization and persistence of these transferred T cells to tumor sites and in the peripheral blood. Thus, biodistribution information of T cells at different time points during therapy would be of use to clinicians. For this reason, modalities to accurately identify and quantify transferred cells in vivo are needed.

Magnetic Particle Imaging (MPI) addresses the need for quantitative tracking of cells in vivo using magnetic nanoparticle tracers. These tracers produce a biorthogonal signal, allowing for unambiguous and direct quantification. If these tracers effectively label T cells, their localization within the patient can be correlated with cell distribution. Using MPI tracers has extreme promise for tracking cell distribution in vivo longitudinally. Our group seeks to improve tracer signal per mass, signal resolution, and tracer surface chemistries for improved cell labeling to progress MPI tracer development for cell tracking purposes.

Particles are first synthesized through a thermal decomposition batch process which yields monodisperse magnetic Fe_3O_4 nanoparticles coated in oleic acid. To be useful in biomedical applications, a polymer coating surrounding the particles is necessary to render them colloidally stable in aqueous media, including cell culture media. Our strategy also introduces unreacted anchor sites near the surface of the particle for further polymer addition and thus additional particle modifications. Polymer choice will provide modularity in particle coating, allowing us to tune their surface chemistry and charge.

We have demonstrated the ability to coat tracers in poly(ethylene glycol) (PEG) and poly(maleic anhydride-alt-1-octadecene) (PMAO), demonstrate the colloidal stability of these particles in several solvents for multiple days, the ability to attach additional ligands onto the surface of the nanoparticles, modify nanoparticle surface charge, attach additional moieties to the hydrodynamic surface of the nanoparticle, and the ability to incubate primary T cells with these tracers in vitro to promote interaction and internalization. We are actively exploring several moieties to further modify the surface of the particle.

Future work will conjugate reactive groups to the nanoparticle surface and incubate functionalized SPIONs with cells of interest to demonstrate improved labeling capabilities. After, these labeled T cells will be introduced to an in vivo model and particle biodistribution will be assessed.

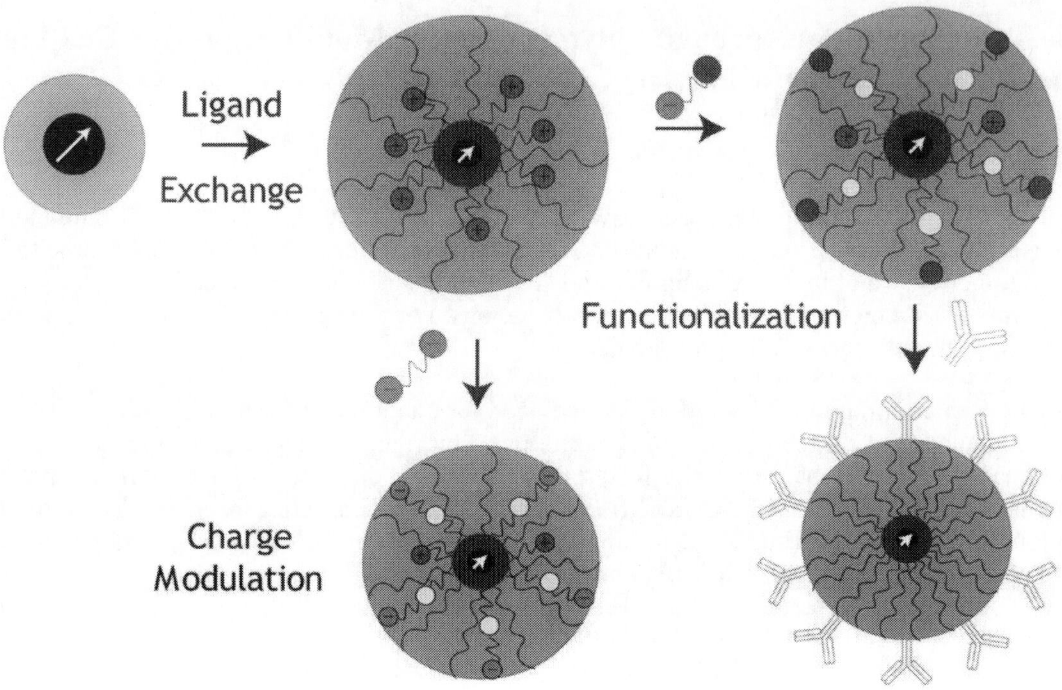
Ligand
Exchange
Functionalization
Charge
Modulation

136f Preparation of Low-Density Cannabidiol-Loaded Nanoparticles and Development of an *in Vitro* Release Assay

Nicholas Caggiano, Brian K. Wilson, Rodney Priestley, Robert K. Prud'homme

Nanoparticle encapsulation is an attractive approach to improve the oral bioavailability of hydrophobic therapeutics. The high specific surface area of nanoparticle formulations, combined with the thermodynamically driven increased solubility of an amorphous drug core, promotes rapid drug dissolution. However, the physicochemical properties of the hydrophobic therapeutic can present obstacles to *in vitro* characterization of nanoparticle formulations. Namely, drugs with low density and high membrane binding affinity frustrate traditional analytical methods to monitor release kinetics from nanoparticles. In this work, cannabidiol (CBD) was encapsulated into nanoparticles 180 nm in diameter with low polydispersity and high drug loading via Flash NanoPrecipitation (FNP), a scalable self-assembly process. Hydroxypropyl methylcellulose acetate succinate (HPMCAS) and lecithin were employed as amphiphilic particle stabilizers during the FNP process. However, the low density and high membrane binding affinity of the amorphous CBD nanoparticle core prevented the characterization of *in vitro* release kinetics by conventional methods. Released CBD could not be separated from intact nanoparticles by filtration or centrifugation. To address this challenge, an alternative approach is described to co-encapsulate 6 nm hydrophobic Fe_3O_4 colloids with CBD during FNP. The Fe_3O_4 colloids were added at 33% by mass (approximately 20% by volume) to increase the density of the nanoparticles, and the resulting particles were 160 nm (CBD-lecithin-Fe_3O_4) and 280 nm (CBD-HPMCAS-Fe_3O_4). This densification enabled centrifugal separation of dissolved (released) CBD from unreleased CBD during the *in vitro* assay while avoiding the losses associated with a filtration step. The resulting nanoparticle formulations provided more rapid and complete *in vitro* dissolution kinetics than bulk CBD, representing a six-fold improvement in dissolution compared to crystalline CBD. The co-encapsulation of high-density Fe_3O_4 colloids to enable separation of nanoparticles from release media is a novel approach to measuring *in vitro* release kinetics of nano-encapsulated low-density, hydrophobic drug molecules.

136g Toxicity Studies of Chiral Nanoparticles *in Vitro* for Medical Applications

Macayla Caso, Michael G. Benton, Kevin McPeak

Chiral nanoparticles have gained a lot of interest because of their unique chiroptical properties. Chirality is defined as an objects ability to not be superimposed over its mirror image. Chirality is commonly seen in nature and has been expanded to nanofabrication. Common fabrication methods of displaying chirality are fabricating geometrically chiral nanoparticles or attaching chiral ligands to surfaces. Objects that have opposite chirality are known as enantiomers, which are the same in all physical properties except for how they interact with the polarized light. Chiroptical effects are often very small, but when a chiral plasmonic nanoparticle is fabricated the chiroptical effects are magnified [1]. This gives greater appeal to the fabrication of chiral plasmonic nanoparticles. The most common method to get chiral particles is with adhesion of chiral ligands to the surface achiral nanoparticles. Attaching these ligands impart their chirality to the surface while providing thermodynamic stability to the nanoparticles [2]. Nanoparticles that are structurally chiral without ligand assistance are often not completely chiral, holding at least one plane of symmetry [3]. In this study chiral nanopyramids, nanoparticles that hold no lines of symmetry, are explored for their cytotoxicity in retinal pigment epithelial cells (ARPE-19). Left and right enantiomers of these chiral nanopyramids are tested along with achiral nanopyramids to show the explore how the chirality changes cellular interaction of these nanoparticles. The cytotoxic effects of size, dosage and surface modifications are also explored. Further cellular damage caused by the nanopyramids is explored with modified cells that fluoresce when DNA damage is experienced. With this study toxicity and DNA damage of achiral and chiral nanopyramids is explored to for medical application limits.

References:

[1]: V. K. Valev, J. J. Baumberg, C. Sibilia and T. Verbiest, *Advanced Materials*, 2013, **25**, 2517-2534.

[2]: W. Ma, L. Xu, A. F. de Moura, X. Wu, H. Kuang, C. Xu and N. A. Kotov, *Chemical Reviews*, 2017, **117**, 8041-8093.

[3]: K. M. McPeak, C. D. Van Engers, M. Blome, J. H. Park, S. Burger, M. A. Gosálvez, A. Faridi, Y. R. Ries, A. Sahu and D. J. Norris, *Nano Letters*, 2014, **14**, 2934-2940.

205a Computationally Guided Design of Ferrite Nanoparticles for Magnetic Inductive Heating

Venkata Rohit Punyapu, Rachel Getman

Identifying alternative methods to efficiently heat catalysts is of interest for both operational and economic reasons. Current technology relies on heating via convection with reactant and product fluids. An alternative promising method is Magnetic Inductive Heating (MIH), where magnetic potential energy is converted into thermal energy to increase the temperature of the system. MIH provides rapid localized heating, efficient heat delivery, and tunable rate of heating, which could be beneficial for catalysts. This form of heating is observed in Magnetically Inductive Materials (MIM), a subset of ferromagnetic materials, where the material's magnetic properties under an alternating external magnetic field create an energy hysteresis to generate heat. A key intrinsic magnetic property of MIMs is Magnetic Anisotropy (MA), which is the MIM's dependence of magnetic moment on the direction of an external magnetic field. In this study, we aim to learn how to control MA by tuning MIM composition. We are specifically interested in ferrites, i.e., 3d metal oxides based on magnetite (which has the chemical formula Fe_3O_4). Ferrites are promising materials as MIM due to their lower cost, durability, and magnetic properties that can be tuned by changing composition through doping. However, at present, the relationship between composition and inductive properties is unknown due to the large composition space. Experimentally, testing the various doped ferrites is time consuming and not tractable; alternatively, it can be accelerated using computational screening. Therefore, in this work, we aim to identify the relationship between ferrite composition and MA using Density Functional Theory (DFT). Specifically, we consider materials with the formulae $M_xFe_{(3-x)}O_4$ ($0 \leq x \leq 1$), where M can be any combination of Mn, Ni, Co, Cu, and Zn. Substitution into octahedral vs. tetrahedral sites and number of substitutions are considered using a high-throughput screening approach. Our results suggest that MA can be tuned by varying ferrite composition. We discuss how these materials can be used as catalysts for the Reverse Water-Gas shift reaction.

205b Mixing Immiscible Elements to Create a Library of Single Phase Ceramic Nanoparticles By Continuous Flame Synthesis

Shuo Liu, Chaochao Dun, Jilun Wei, Jeffrey J. Urban, Mark Swihart

Ceramic solid solution nanoparticles incorporating two or more cations in a single homogeneous phase can have tunable features (*e.g.*, active sites, vacancies, electronic structure) that can yield better performance than unary oxides in many fields, including catalysis. However, most reported ceramic solid solution materials combine similar elements (*i.e.* those with similar crystal structure, atomic diameter, valence, and electronegativity), such as alkaline-earth elements that form thermodynamically stable phases as bulk materials. This limitation to miscible elements dramatically limits the compositional space for new materials discovery. For example, incorporation of nickel into oxide phases that are used as catalyst supports would be desirable, because nickel is an inexpensive metal with broad catalytic activity for commercially important reactions. However, Ni-containing ceramic solid solutions and related phase diagrams have rarely been reported, because the conventional wet-chemistry methods such as co-precipitation performed at low temperature over long reaction times lead to separation of immiscible phases. In recent years, some nonequilibrium synthesis methods employing high temperature and short reaction time have been developed for mixing immiscible elements into a single ceramic or alloy phase, such as carbo-thermal shock, laser ablation, and spark discharge, but these methods often face barriers of scalability due to complexity, low yield, and high cost.

Flame aerosol processing is the most common method for scalable production of low-cost nanomaterials. Here, we reported a continuous, low-cost, and scalable flame synthesis method to overcome immiscibility and create a library of binary ceramic solid solutions with applications in catalysis. As shown in the figure, an aqueous precursor containing two metal salts is delivered to the flame reactor and atomized into microdroplets within which evaporation and reaction drives formation of nanoparticles. The reactor resistance time (~0.05 s) is much shorter than the time required for phase separation by solid-state diffusion, so the initial state of uniform composition in the droplet precursor can be retained in the product. Afterward, rapid quenching with diluting nitrogen prevents phase separation, finally forming a single binary ceramic solid solution. Here, we demonstrated the generality of this approach by mixing various elements including transition metals, noble metals, alkaline-earth elements, and others.

As a prototypical application, an amorphous Ni-Al ceramic solid solution is used to create a catalyst for dry reforming of methane. In this case, the initial metastable single phase undergoes nickel exsolution in a reducing atmosphere to create a nickel-decorated alumina catalyst with excellent catalytic activity and stability. Moreover, the droplet-to-particle conversion process by which the catalyst is formed results in a hollow micro- or nano-sphere morphology. Addition of a surfactant to the precursor solution allows templating of pores in the hollow shell which increase the surface area and make the interior of the shell accessible. The final hollow, porous, nickel-decorated alumina catalyst maintains over 90% CO_2 and CH_4 conversion for hundreds of hours at 800 °C when used as a dry reforming catalyst.

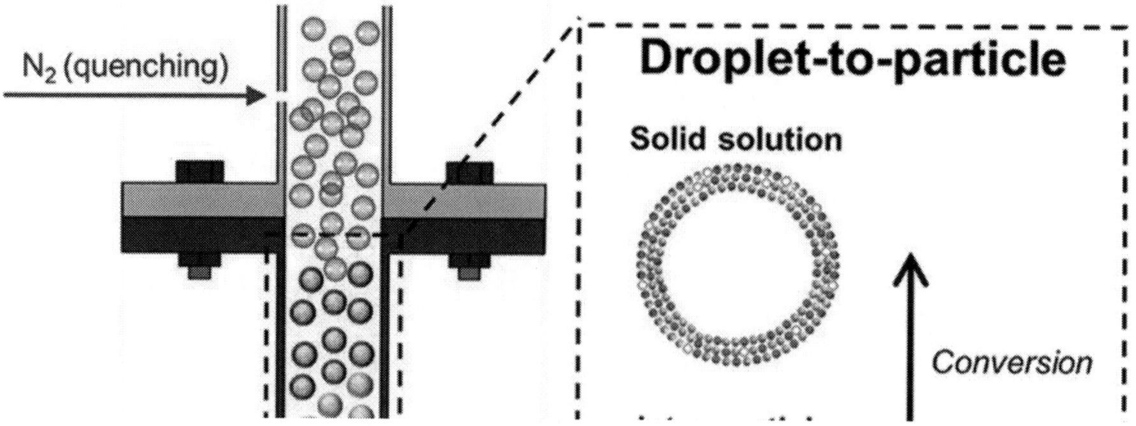

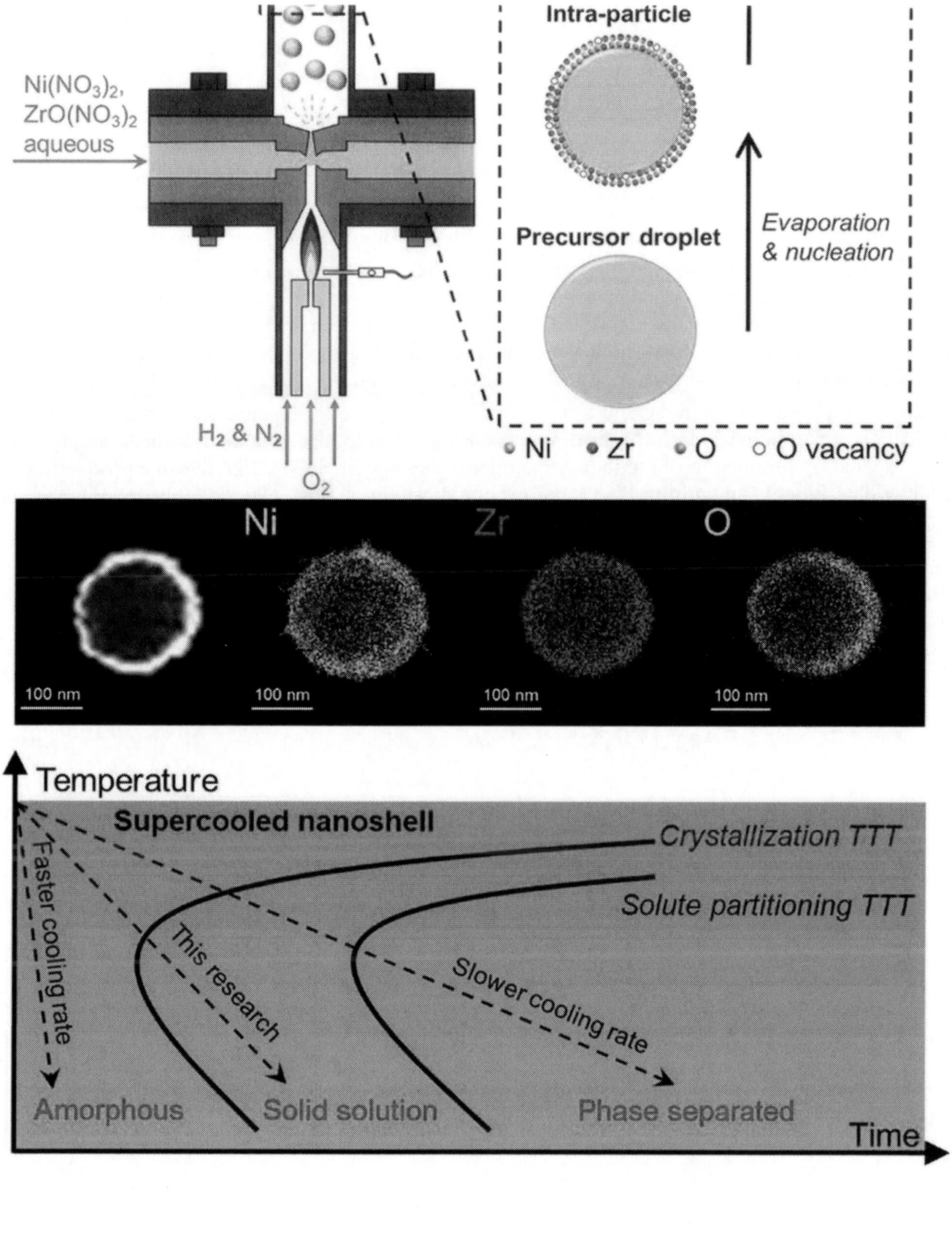
Ni(NO3)2,
ZrO(NO3)2
aqueous
H2 & N2
O2
Intra-particle
Precursor droplet
Evaporation
& nucleation
Ni
Zr
O
O vacancy
Ni
Zr
O
100 nm
100 nm
100 nm
100 nm
Temperature
Supercooled nanoshell
Crystallization TTT
Solute partitioning TTT
Faster cooling rate
This research
Slower cooling rate
Amorphous
Solid solution
Phase separated
Time

205c Preparation of Large-Pore-Volume Al_2O_3 Microspheres As Catalyst Support Via Microfluidics

Huilin Yi, Yujun Wang, Guangsheng Luo

Mesoporous alumina (Al_2O_3) is regarded as a potential candidate in catalysts, the textural properties of which affect the structure and performance of heterogeneous catalysts greatly. Herein, we report a systematic investigation of the preparation of large-pore-volume Al_2O_3 microspheres using microfluidics. The pore expansion mechanism of synthesis system with methylcellulose (MC) as the template and gelation initiator and an organic solvent as the swelling agent and carrier medium was investigated experimentally and theoretically. The coaxial microchannel was subjected to molding; this process significantly influenced the morphology and textural properties owing to the internal circulation during droplet formation, by which the obtained microspheres exhibited an optimal specific surface area of 403 m^2/g and a pore volume of 1.85 cm^3/g. To meet the demand for industrial application, a parallelized microfluidic device with improved ladder–tree distributors were developed for the scaled-up production of monodisperse microspheres with a CV of 2.58% in a production rate of 500 g/day. Finally, an optimized CO_2 conversion of 87.0% at 300 °C under atmospheric pressure was realized owing to the large pore volume and uniform Ni dispersion of Ni/Al_2O_3 microsphere catalysts. This work provides a feasible strategy of preparing Al_2O_3 microspheres with large pore volume as promising catalyst support, and have potential in the extensive industrial application for the preparation and molding of mesopore materials.

205d Dial-a-Material: Precise Manufacturing Technology for Bimetallic and Monometallic Nanoparticles

Bruno Pinho, Laura Torrente-Murciano

In this work, we pioneer an autonomous, self-regulated manufacturing flow platform for the synthesis of targeted bi-metallic nanoparticles with dial-a-particle capabilities up to 100 nm sizes. We showcase the technology for monometallic (Ag) and bimetallic core-shell (Ag@Pt) nanoparticles to demonstrate its potential for the distributed manufacturing of materials on demand with minor human-intervention. This technological platform offers: (i) materials on-demand (wide range of sizes), (ii) distributed manufacturing, (iii) minor human-intervention, (iv) controllable multi-stage synthetic protocols, (v) real-time multi-point particle characterization (v) modular flexibility, and (vi) precision. The novelty of this work lies in the modular approach of the platform equipped with low-cost spectroscopy that enables fast (within a few seconds) monitoring at multiple points of the synthesis. Size tuneability is successfully achieved in the absence of organic capping ligands, using early growth information to mechanistically predict the resulting particle properties. We demonstrate that access to early-stage data in long synthetic routes (> 10 minutes) is vital to ensure a smooth and stable operation with minor user intervention. This work goes beyond the previous state-of-the-art of automated manufacturing approaches and rapid characterization [4], limited until now to single-step and rapid (~minute per iteration) syntheses [3]. It also advances our previous work on the continuous synthesis of seed-mediated growth of silver citrate-based nanoparticles with sizes ranging from 4 to 80 nm using multi-stage flow systems [5] where lack of automation leads to an off-target average inconsistency of ±13% (>100 syntheses). Indeed, herein, we demonstrate control not only within the variable reaction parameters (e.g. concentration, pH, time per stage, micromixing, etc) but also uncontrollable parameters that detrimentally affect the reproducibility and repeatability of the system such as precursor ageing, feedstock inconsistency, flow fluctuations, fouling, speciation, metal traces, etc. [2, 5].

As a result, this work will close the gap between nanomaterial research [1] and their near-future industrial deployment [2] opening the door to exciting opportunities for nanoparticles in a wide range of applications such as catalysis, energy materials, bio-medicine, and bio-imaging.

References:

[1] L. Liu, A. Corma, *Chemical Reviews* **2018**, 118, 4981; U. Aslam, S. Chavez, S. Linic, *Nature Nanotechnology* **2017**, 12, 1000; N. G. Bastús, F. Merkoçi, J. Piella, V. Puntes, *Chemistry of Materials* **2014**, 26, 2836.
[2] Y. Gao, B. Pinho, L. Torrente-Murciano, *Current Opinion in Chemical Engineering* **2020**, 29, 26.
[3] S. Li, R. W. Baker, I. Lignos, Z. Yang, S. Stavrakis, P. D. Howes, A. J. deMello, *Molecular Systems Design andamp; Engineering* **2020**; L. Bezinge, R. M. Maceiczyk, I. Lignos, M. V. Kovalenko, A. J. deMello, *ACS Applied Materials andamp; Interfaces* **2018**, 10, 18869.
[4] R. W. Epps, K. C. Felton, C. W. Coley, M. Abolhasani, *Lab on a Chip* **2017**, 17, 4040; K. Abdel-Latif, R. W. Epps, C. B. Kerr, C. M. Papa, F. N. Castellano, M. Abolhasani, *Advanced Functional Materials* **2019**, 29, 1900712; H. Bolze, P. Erfle, J. Riewe, H. Bunjes, A. Dietzel, T. P. Burg, *Micromachines (Basel)* **2019**, 10, 179.
[5] B. Pinho, L. Torrente-Murciano, *Reaction Chemistry andamp; Engineering* **2020**, 5, 342.
[6] B. Pinho, L. Torrente-Murciano, *Advanced Energy Materials* **2021**, 11, 2100918.

Development of green roofing tiles with the use of novel photocatalytic and reflective TiO_2 nanoparticles

M. Kouroutzi[1], A. Stratidakis[2], M. Kermenidou[1], C. Gabriel[1], M. Perikli[3], E. Feleki[4], K. Karatisoglou[5], V. Kakoutopoulos[5], A. Kothalis[5], S. Karakitsios[1,6], D. A. Sarigiannis[1,2,6*]

[1] Aristotle University of Thessaloniki, Department of Chemical Engineering, Environmental Engineering Laboratory, University Campus, Thessaloniki 54124, Greece

[2] University School for Advanced Study (IUSS), Science, Technology and Society Department, Environmental Health Engineering, Piazza della Vittoria 15, Pavia 27100, Italy

[3]Modern Analytics, Thessaloniki, Greece

[4] Laboratory of Heat Transfer and Environmental Engineering, Aristotle University, Box 483, 54124 Thessaloniki, Greece

[5] KEBE-Northern Greece Ceramics S.A., Nea Santa, Kilkis 61100, Greece

[6] HERACLES Research Center on the Exposome and Health, Center for Interdisciplinary Research and Innovation, Balkan Center, Bldg. B, 10th km Thessaloniki-Thermi Road, 57001, Greece

http://www.enve-lab.eu

**** Corresponding Author***

Professor Dr Dimosthenis A. Sarigiannis

Email: denis@eng.auth.gr

Keywords: *TiO_2 nanoparticles, photocatalytic properties, photoreflectance, green tiles, characterization techniques*

1. Introduction

Given the great significance of a roof in the energy efficiency and heat conditions of a room, solutions for the design of the so-called "cool roof" are very important to be developed. Cool roofs are capable of enhancing the reflectance properties of a building, thus absorb a minimum amount of heat. This way, the roof is able to remain cool and maintain the building's internal temperature stable, by reflecting the greatest percentage of the sunlight [1].

This can be achieved by adding a thin film coating over the roof tiles. Using this technique, we maximize heat emission and minimize the absorption of solar radiation while also maintaining a steady and cool temperature inside the building. Titanium dioxide (TiO_2) is a largely available and low-cost chemical which exhibits high photocatalytic and photoreflective activity, while being non-hazardous to human and environmental health. It is also well-known for its ability to degrade organic pollutants [4]. It occurs as two important polymorphs, rutile and anatase. Both polymorphs demonstrate differentations among their properties, including their photoreflective and photocatalytic activity. Anatase is metastable, exhibiting high photoreactivity and high photoreflectance of the long wave ultraviolet (UVA) and visible light. On the contrary, rutile is more stable and absorbs violet visible light [5].

This research study was conducted towards the research project entitled "Green Tile evelopment-KERAMI" in collaboration with "KEBE-Northern Greece Ceramics", aiming to develop a novel "green" roof tile with photoreflective and photocatalytic properties.

2. Materials and Methods

The clay roofing tiles were manufactured and provided by the "KEBE-Northern Greece Ceramics". Three types of TiO_2 materials were purchased from Sigma-Aldrich (USA), including anatase, rutile and a mixture of rutile and anatase nanoparticles. Polyethylene glycol (PEG) (MW: 600) was purchased from Baker, Germany.

Novel TiO_2 nanoparticles were developed along with the addition of polyethylene glycol (PEG), which was added as a surface active material, to increase the catalytic surface of the nanoparticles. For this reason, deionized water was mixed with 2.5 % w/v of one of each of the three types of TiO_2 and 0.208 mass % of PEG (MW: 600), while inserted into an ultrasonic bath (50 min). For the characterization of the nanoparticles, all prepared samples were subsequently dried at 100°C and 290°C. At 290°C, PEG degrades and leaves a mesoporous photocatalytic structure. TiO_2 nanoparticles without PEG have also been developed as control samples, following the same procedure. In total, 12 samples with and without PEG have been developed. For the preparation of the final "green" clay roofing tiles, aqueous suspensions of all nanoparticles were sprayed onto the surface of the roof tiles and formed a coating with photocatalytic and photoreflective properties.

XRD analysis was performed for the determination of the phase compositions of the purchased nanoparticles. Experiments were conducted on a Brucker D8 Advance diffractometer operating in the reflection mode with Cu Ka radiation in the 2θ interval of 10-80 °C.

The particle size distribution of the nanoparticles was estimated by a laser diffraction technique using a Malvern Instruments, zeta-nano series, Nano ZS. In order to obtain clear results in particle size distribution analysis, the aqueous suspensions of all examined samples underwent ultrasonic action for 10 minutes, prior to the measurements.

BET method was used for the estimation of the specific surface area of the nanoparticles. Experiments were conducted using a Micromeritics TriStar instrument, by low temperature nitrogen sorption.

The photocatalytic activity of the samples was evaluated by monitoring the decomposition of NO gas. The experimental set-up was developed by the research team and consisted by the reactor, a UV source, a nitric oxide (NO) supply and flow rate valves. NO gas with synthetic air were inserted into the layout, into the chamber where the coated tiles were placed and subsequently irradiated with UV light. The decrease in concentration of pollutants was recorded at predetermined time intervals.

The reflectance of the samples was measured by a spectrometer IFS-113V made from Bruker Physik AG, Silberstreifen, 7512 Karlsruhe-Rheinstettent/B.R. Deutschland. During measurements, an integrated sphere was used for the collection of both specular and diffuse reflectance. The spectrum of the studied wavelength of reflectance ranged from 200 to 900 nm.

3. Results and Discussion

The diffraction patterns exhibited strong diffraction peaks corresponding to anatase and rutile phases. The percentage of both crystalline phases in the purchased TiO_2 materials was determined by an X-Ray diffraction meter. The quantitative analysis for these specimens provided anatase and rutile analogues. For samples prepared from TiO_2 anatase, the analogue was approximately 90% for anatase and 10% for rutile. For samples prepared from the mixture of anatase and rutile TiO_2, the analogue was approximately 80% for anatase and 20% for rutile.

Results obtained from the particle size distribution analysis exhibited that particle sizes of all samples range within the fraction of 70 to 110 nm, with the sizes of PEGylated nanoparticles exhibiting slightly greater sizes.

For the estimation of the specific surface area of the newly developed nanoparticles, all samples were dried at 290°C. According to the results obtained by the BET analysis, the addition of the polymer changes the specific surface area of the samples. More specifically, TiO_2/PEG nanoparticles exhibit greater surface areas (91.5 m^2/g) than the TiO_2 nanoparticles without PEG (77.7 m^2/g). For the mixture of rutile and anatase TiO_2/PEG nanoparticles, the specific surface area was 58.9 m^2/g, while for the same type of nanoparticles without PEG, the respective surface area was 49.9 m^2/g. The same trend was also shown by rutile TiO_2/PEG and TiO_2 rutile nanoparticles. More specifically, rutile TiO_2/PEG (36.6 m^2/g) nanoparticles exhibited a greater surface area comparing to rutile TiO_2 without PEG (31.0 m^2/g), when dried at 100 °C.

In all cases, the addition of PEG lead to an increase in the surface area of the samples, which is attributed to the mesoporous photocatalytic structure created after the degradation of the polymer.

Reflectance measurements demonstrated great results with the reported total reflectance (UV-Vis, visible light and Infrared) exceeding 70%. Additionally, results obtained from the experimental procedure for the estimation of the photocatalytic activity of the samples, exhibited the great photocatalytic activity of the novel nanoparticles in the decomposition of the NO gas.

4. Conclusions

The results reported in this study indicate that the novel TiO_2 nanoparticles, used as a photocatalytic and photoreflective coating onto the surface of roof tiles, exhibit high reflectivity and good photocatalytic activity.

"Green" roof tiles prove to be ideal for the enhancement of environmental protection and reduction of air pollution by decomposing pollutants, while also promoting the decrease of energy consumption, due to reduced use of air conditioning systems, mainting cool temperatures in the interior of buildings occupants and reducing the urban-heat island phenomena.

5. Funding sources

This research study was funded by the Operational Programme Competitiveness, Entrepreneurship and Innovation 2014-2020 (EPAnEK) of the Ministry of Economy and Development, aiming to develop a novel "green" tile which will have photoreflective and photocatalytic properties for use in cool roofs.

References

[1] (Yun, Cho, & Cho, 2018) Yun, Y., Cho, D., & Cho, K. (2018). *Performance evaluation on cool roofs for green remodeling.* Paper presented at the AIP Conference Proceedings.

[2] U.S. Department of Energy, Guidelines for selecting cool roofs. 2010.

[3] Yang, J., Bou-Zeid, E., 2019. Scale dependence of the benefits and efficiency of green and cool roofs. Landscape and Urban Planning. 185, 127-140.

[4] Bahnemann, D., 1999. Environmental Photochemistry, The Handbook of Environmental Chemistry. Springer.

[5] Simons, P. Y., Dachille, F., 1967. The structure of TiO2II, a high-pressure phase of TiO2. Acta Crystallographica. 23, 334-336.

205f Gas Phase Coating of Fluidized Particles: Towards Ton-Scale Production with Nano-Precision

J Ruud Van Ommen

Nanostructured particles (e.g., core-shell nanoparticles or porous micro-particles containing nanoparticles) have high potential in applications such as catalysis and energy storage. However, to come from product design incorporating nanostructuring to actually making such materials at relevant scales is not trivial. Gas-phase coating using techniques such as atomic layer deposition (ALD) and molecular layer deposition (MLD) can be used to provide the surface of a particle with either an ultrathin continuous coating or a decoration of nanoclusters. In this way, we can take benefit from the wide range of chemistries that has been developed over the years for ALD and MLD.

When carried out in a fluidized bed or a pneumatic transport reactor, gas-phase coating is an attractive way of producing nanostructured particles with excellent scale-up potential. Even nanopowders can be fluidized to be processed this way, although they are not present as individual particles but as very dilute agglomerates. Since this often involves cohesive materials, an effort has to be made to obtain proper dispersion of the particle.

I will show some examples of the use of ALD and MLD to powders. I will demonstrate how it can be used to dramatically increase the lifetime of cathode powders for Li-ion batteries and phosphor particles for LED lights. Moreover, the application to various areas of catalysis – photo-, electro-, and thermal catalysis – will be discussed.

205g Flame Aerosol Synthesis of High-Entropy Ceramic Nanoparticles

Shuo Liu, Chaochao Dun, Jilun Wei, Jeffrey J. Urban, Mark Swihart

Tailoring properties by creating mixtures is an age-old method to improve material performance. The main mixing strategy is addition of a small amount of second component into a primary component, like addition of a small amount of carbon into iron to create steel, which greatly increases strength and corrosion resistance. However, this conventional strategy limits the range of possible material types and available composition space. In 2004, scientists discovered that an alloy containing more than five elements in near-equiatomic concentrations can be stabilized in a single phase by its high configurational entropy. This insight opens up a vast compositional space for new materials discovery. It was not until 2015 that the concept of "high-entropy ceramics" was first proposed with the rock-salt structured $(MgCoNiCuZn)O_X$ solid solution oxide reported. As a highly disordered, multicomponent, and often metastable system, high-entropy ceramics can exhibit interesting and useful new properties, like high-temperature stablity, extreme lattice distortions and abundant oxygen vacancies. New and unexpected properties may emerge upon tuning the elemental composition and ratio. Features ranging from catalytic activity to lithium-ion uptake to low thermal conductivity at high temperature can be exploited in energy and environmental applications such as catalysis, sensors, supercapacitors, insulation, and electrode materials.

To date, the brute-force ball milling method has been the most common strategy to fabricate high-entropy ceramics, but this approach has limited ability to control particle size and prevent phase-separation. More recently, bottom-up synthesis methods have been developed, including carbo-thermal shock, laser ablation, and spark discharge, but these methods often face barriers of scalability due to the complex processes, low yield, and highly energy required. Here, we reported a continuous, low-cost, and scalable flame aerosol process to synthesize high-entropy ceramic nanomaterials. In a general synthesis, an aqueous precursor containing multiple metal salts is delivered to a flame reactor and atomized into microdroplets within which evaporation and reaction drives formation of the solid nanoparticles. The reactor resistance time (~0.05 s) is much shorter than the time required for phase separation by solid-state diffusion, so the initial high-entropy well-mixed state produced at high temperature in the reactor can be retained in the product. The rapid quenching with diluting nitrogen prevents phase separation so a single ceramic phase can be obtained. The closed reactor chamber provides a controllable oxidizing or reducing environment that allows further tuning of, for example, oxygen vacancy concentration. We demonstrated the generality of above mechanism by mixing various elements, including transition metals, noble metals, alkaline-earth elements and others. Many different phases, each of uniform well-mixed composition, were produced, such as amorphous, rock-salt, and fluorite structures. In addition, many products form a hollow nanoshell structure directly in the droplet-to-particle process by which the nanostructures form. Explorations of this class of high-entropy ceramic nanoparticles for catalysis and Li-battery applications are ongoing.

205h Filtration of NaCl Aerosols Using Porous Versus Non-Porous Filter Media

Amulya Poudyal

Commercial filter media such as MERV rated filters and HEPA are typically made of glass fibers which are non-porous. These filters use traditional filtration mechanisms such as interception, inertial impaction and diffusion, which involve sticking filtrates to the surface of the fibers. This has been explained by the single fiber efficiency model. These surface filters have a relatively short useful life because they clog with filtrates after brief use and go to a high pressure drop state rendering it useless. To overcome this problem, a novel three dimensional filter media with nanofiber flocks embedded within the microfibrous ACFs matrix was designed. This media is hereinafter referred to as "AU filter media".

The single fiber filtration efficiency model does not account for the role of internal porosity in filtration. The movement of filtrate into the pores of the fibers have not been contemplated before. We hypothesized that the inherent wet pores of hydrophilic ACFs can potentially wick particulates such as NaCl until saturation, thus increasing the dirt holding capacity and therefore prolonging the life of the filter media. In addition, highly porous nanofiber flocks made of Vapor Grown Carbon Fibers (VGCF) were utilized because of the potential of the nanofibers to capture filtrates via wall slip phenomenon. Wall-slip occurs when the diameter of the fiber approaches the mean free path of air. This causes an increase in filtration efficiency without the anticipated pressure drop penalty.

Preliminary filtration experiments were conducted for NaCl aerosol filtration. The results showed that the initial filtration efficiency and quality factor improved with nanofiber flock inclusion. The EDS spectrum verified the preferential loading of NaCl into VGCF flocks. This was attributed to the wall-slip occurring over nanofiber flocks. The decrease in BET surface area and total pore volume of ACFs, after loading it with NaCl, indicated the transfer of NaCl into the wet pores of ACFs. More importantly, the dirt holding capacity of the AU media was found to be higher than MERV 16 and HEPA at both low and high RH conditions. In conclusion, this novel filter media shows promising results for long term and high dirt loading filtration applications due to the dual effect of wall-slip occurring over nanofiber flocks and intra-fiber NaCl storage in ACFs.

269b 'Ringing' Breast Tumors with Ligand-Specific X-Ray Contrast Agents and Spectral CT

Kalyan Ramesh, Alice Truong, Mary Ruskowski, Manos Gkikas

X-ray-computed tomography (CT) is considered one of the powerful diagnostic techniques for preclinical and clinical studies due to deep tissue penetration, high resolution, cost effectiveness, rapid scan times and patient comfort. Non-invasive disease detection with pronounced signal enhancement at the site of injury is a great challenge in medicine and allows for condition estimation in early stages. Designing metallic nanoparticle (NP) contrast probes for spectral CT imaging that can be statistically discriminated from biological fluids, tissue, and bone (a property that is impossible to achieve with conventional CT), and are empowered by *molecular recognition*, aligns towards that direction. Our Team (UML-Chemistry / UMMS-Radiology) is developing ***X-ray contrast probes based on molecular recognition for breast cancer cells***, aiming to achieve early disease diagnosis and enhanced detection at low NP dosages. ***Among our recent findings is the ability to detect breast tumors in vivo for up to 21 days with <0.5 mg of Au***. Our approach combines ***NP targeting specificity*** with ***dual-energy CT or photon-counting spectral CT (PCD-CT)***, a state-of-the-art form of spectral CT that allows for visualization of tissue composition based on intensity of signal, which is dictated by *atomic number and density of contrast materials*, and thus is enhanced by metallic NPs. Au offers significant advantages towards materials differentiation since its K-edge is 81 vs. 4 keV of Ca, while its density is 19.6 vs. 1.6 g/mL of Ca. Since X-ray absorption is highly dependent on K-edge and density of materials, ***we synthesize Au-based nano-contrast agents with high affinity for breast cancer cells and CT contrast comparable/higher to FDA-approved iodinated molecules*** *at equal metal basis*. Our materials are tested *in vitro* as well as *in vivo* with relevant animal models. The most recent findings on animal studies will be highlighted.

269c Platinum Nanoparticles Encapsulated within PLGA, Treatment for TNBC, an *in Vitro* and *In Vivo* study

Aida López Ruiz, Kathleen McEnnis

Nanotechnology is a multidisciplinary field that has many different applications including cancer treatment and detection. Noble metal nanoparticles such as gold, silver or palladium, have shown major potential in the field of medicine and pharmaceutics. In particular, platinum nanoparticles (PtNPs) have been reported with promising anticancer results. The activity of nanoparticles is based on their small size and high surface area; these properties allow them to penetrate biofilms as well as influence intracellular mechanisms. Recent studies indicate that Pt NPs can be used as a therapy for cancer showing a limited toxicity to healthy cells. Hence, in this work, we developed a new treatment option for triple negative breast cancer (TNBC) based on Pt NPs. TNBC is a type of breast cancer that lacks the three common receptors used for immunotherapy and with a tendency to metastasize faster. Therefore, TNBC has a worst prognosis and often chemotherapy is the only treatment option. However, chemotherapy is toxic, has many side effects and some cancers can even develop resistance, therefore, new treatment options are needed to combat TNBC.

Drug delivery systems are technologies that help enhance circulation time or targeting at the tumor site. Therefore, to avoid accumulation in the liver or spleen a drug delivery system was proposed. Poly(lactide-co-glycolide) (PLGA) particles were synthesized as a delivery system for PtNPs. PLGA is a copolymer of poly(lactic acid) (PLA) and poly(glycolic acid) (PGA), two monomers that are metabolized by the body, making PLGA toxicity minimal. The encapsulation of PtNPs within PLGA nanoparticles was performed by nanoemulsion. Moreover, to enhance the circulation lifetime of the particles, surface modifications with polymers such as polyethylene glycol (PEG) can be performed. Surface modifications with PEG are a commonly used procedure to enhance circulation-time. Furthermore, active targeting by ligands of specific biomarkers, monoclonal antibodies, peptides, and aptamers provides the most effective therapy. TNBC is defined by the lack of overexpression of the three common receptors used to classify breast cancer. Therefore, there are very few available targets. Epidermal growth factor (EGF) is a receptor that promotes cell proliferation migration and angiogenesis (development of new blood vessels). EGFR is overexpressed in many tumors including breast cancer. Therefore, many strategies based on antibodies have been studied to block EGFR. The anti-EGFR monoclonal antibody, Cetuximab, has shown promise for targeting metastatic TNBC.

In the present study, PtNPs were synthesized, and the anticancer activity was tested *in vitro* against TNBC, showing a remarkable cell death. As the major drawback of traditional chemotherapy is the toxicity, cell viability experiments were also performed on fibroblasts. The results showed no apparent toxicity of Pt NPs for healthy cells in contrast to cisplatin (a common chemotherapeutic drug). Encapsulation of PtNPs within PLGA was performed by nanoemlusion with a 60% loading percentage. Surface modifications with PEG and EGFR were performed to increase circulation time, and achieve a higher anticancer effect. Different combinations of PEG/EGFR were synthesized and tested *in vitro* with TNBC, achieving a better efficacy using 50% PEG and 50% EGFR on particles surface. Based on the promising results achieved *in vitro*, the effect of bare PtNPs and 50% EGFR PtNPs were tested *in vivo* with mice. In summary, this work provides evidence of a potential new treatment for TNBC based on Pt NPs with no apparent toxicity for healthy cells, overcoming one of the major drawbacks of current treatments.

269d A Personalized Approach Towards the Delivery of Itraconazole Nanocrystals Onto Contact Lenses Using Inkjet Printing

Carolin Tetyczka, Kira Brisberger, Martin Reiser, Manuel Zettl, Ramona Jeitler, Christina Winter, Dagmar Kolb, Gerd Leitinger, Martin Spörk, Eva Roblegg

Ocular administration of drugs is usually performed via eye drops or ointments. This most often increases the production of tear fluid, which causes the drug to be easily flushed out of the eye, resulting in a very short residence time and consequently low bioavailability. Approaches to extend the residence time include mucoadhesive ophthalmic ointments/films, eye drops that form a viscous gel upon administration, and contact lenses. Drug-loaded contact lenses are of particular interest as they can increase the residence time of drugs from 1-2 min to more than 30 min, which results in a 50% increased ocular bioavailability [1]. A disadvantage, however, is that drug-loaded contact lenses may impair the visibility of patients. A promising method to counteract this problem is inkjet printing. Inkjet printing enables the targeted printing of picoliter droplets onto substrates with high precision, allowing the production of personalized dosage forms [2]. A decisive factor for the printability of the ink solution is the solubility of the drug in the respective solvent. In order to increase the solubility of poorly soluble drugs and thus achieve therapeutic doses, particle size reduction of the bulk material into the nanometer range can be performed. The obtained nano-formulation can then be printed onto the lens in a targeted manner [3].

This study aims at demonstrating the selective printability of a nano-formulation based on the model drug Itraconazole (ITZ) on commercially available soft hydrogel 1-day contact lenses using inkjet printing. First, ITZ nanocrystals were prepared using wet media milling, optimized via Design of Experiments (DoE) and carefully characterized in terms of particle size, shape, solid-state behavior, and saturation solubility. In a second step, the printing process was adjusted regarding optimum firing frequency and pulse shape. Effects of the jetting process on particle size, size distribution, crystallinity and content uniformity were evaluated. Finally, printing trials on contact lenses were conducted.

As a result of DoE analysis and optimization, nanocrystals with a size of approximately 200 nm, a narrow particle size distribution and an almost spherical shape were obtained. Compared to the bulk material, the solubility was significantly increased and no changes in the solid-state behavior occurred. Inkjet printing into vials did not negatively affect the characteristics of the nanocrystals. In addition, multiple layers (i.e., up to 10) could be printed on the contact lenses using the configured ring pattern. The area of the pupil was selectively omitted, thereby avoiding any visual impairment. This highlights that printing on contact lenses is a promising application for ophthalmic drug delivery, allowing personalized dosing without affecting visibility.

REFERENCES

1. F.A. Maulvi, D.T. Desai, K.H. Shetty, D.O. Shah, M.D.P. Willcox, Advances and challenges in the nanoparticles-laden contact lenses for ocular drug delivery, Int. J. Pharm. 608 121090 (2021).
2. W.H. Chou, A. Gamboa, J.O. Morales, Inkjet printing of small molecules, biologics, and nanoparticles, Int. J. Pharm. 600 120462 (2021).
3. W.S. Cheow, T.Y. Kiew, K. Hadinoto, Combining inkjet printing and amorphous nanonization to prepare personalized dosage forms of poorly-soluble drugs, Eur. J. Pharm. Biopharm. 96 314–321 (2015).

269e Indocyanine Green J-Aggregate and Anticancer Drug Loaded Polymersomes for Photoacoustic Image-Guided Drug Delivery

Ceren Atila Dincer, Mohammed Kawelah, Sangheon Han, Thomas Truskett, Konstantin V. Sokolov, Keith P. Johnston

Theranostic methods to image and treat cancer have the potential to have a large impact on human health. Photoacoustic imaging (PAI) using near-infrared (NIR) contrast agents allow deep tissue imaging, for example of breast and ovarian cancer. Herein, we encapsulate indocyanine green J-aggregates (ICGJ) and drugs into polymersomes to achieve sufficient blood circulation and targeting of tumor cells with monoclonal antibodies on the surface. The polymersomes were synthesized via self-assembly of poly(lactide-co-glycolide-*b*-polyethylene glycol) (PLGA-*b*-PEG) amphiphilic block copolymer using a double emulsion solvent evaporation method. ICGJ encapsulated polymersomes exhibited sizes of 80-90nm and strong NIR absorbance at 890nm. The polymersomes were stable in neutral (pH 7.4) and acidic (pH 5.5) 100% fetal bovine serum (FBS) for 24 hours enabling photoacoustic cancer imaging. A hydrophobic anticancer drug paclitaxel was loaded along with ICGJ for image-guided therapy to monitor the drug release.

269f Eradication of Antibiotic-Resistant Biofilms By Hyperthermia Using Superparamagnetic Iron Oxide Nanoparticle Films

Shaquib Rahman Ansari, Georgios Sotiriou, Alexandra Teleki

Implant-associated infections (IAIs) are a common complication associated with surgery failures. They are characterized by biofilm formation, which adhere to the surfaces of medical devices and are resistant against conventional antibiotic therapies. Therefore, there is a need to identify novel treatment strategies capable of killing established biofilms and/or prevent their formation (1). In this study, magnetic hyperthermia (MH) induced by superparamagnetic iron oxide nanoparticles (SPIONs) is used for biofilm eradication. SPIONs generate heat upon exposure to an alternating magnetic field (AMF), which can be used alone or in combination with conventional antibiotic therapies to induce biofilm eradication. The thermogenic bactericidal approach can facilitate an increased uptake of antibiotics to the bacterial cells and also overcome the limitations of chemical biocides that are susceptible to the differences in metabolic states of the bacteria. Here, implant surface coated with doped-SPION films were used to study the effect of MH on *Staphylococcus aureus* biofilm formation alone and in combination with vancomycin.

Manganese ferrite nanoparticles ($Mn_{0.5}Fe_{2.5}O_4$) of size 18 nm were in situ deposited during their production onto 10 mm x 10 mm silicon substrates using flame spray pyrolysis (2). These nanoparticles show higher saturation magnetization (67 emu g^{-1}) and magnetic hyperthermia (ΔT = 164 °C in 10 sec) than undoped SPIONs (3). Multiple Mn ferrite layers were deposited by encasing each layer in a polymer before the consecutive deposition. The nanoparticle composite film had a coating density of 1.3 mg cm^{-2} and each nanoparticle film was 1.6 μm thick.

This study demonstrates that SPIONs can be successfully deposited on substrates and increase the surface temperature. The hyperthermia properties of the SPION film in combination with conventional antibiotics will be used to eradicate the biofilms of clinically relevant bacteria.

Acknowledgement: The authors would like to acknowledge the financial support from the Science for Life Laboratory. This project has received funding from the European Research Council (ERC) under the European Union's Horizon 2020 research and innovation programme (grant agreement No. 101002582)

References:

1. Alumutairi L, Yu B, Filka M, Nayfach J, Kim M-H. Mild magnetic nanoparticle hyperthermia enhances the susceptibility of Staphylococcus aureus biofilm to antibiotics. Int J Hyperth. 2020 Jan 1;37(1):66–75.

2. Sotiriou GA, Blattmann CO, Pratsinis SE. Flexible, Multifunctional, Magnetically Actuated Nanocomposite Films. Adv Funct Mater. 2013 Jan 7;23(1):34–41.

3. Ansari SR, Hempel N-J, Asad S, Svedlindh P, Bergström CAS, Löbmann K, et al. Hyperthermia-induced in situ drug amorphization by superparamagnetic nanoparticles in oral dosage forms. ChemRxiv. Cambridge: Cambridge Open Engage; 2022; This content is a preprint and has not been peer-reviewed.

269g Thermal Decomposition Synthesis of Iron Oxide Nanoparticles in a Precision Machined Reactor with Enhanced Gas-Liquid Mass Transfer for Magnetic Particle Imaging (MPI)

Ambar C. Velazquez-Albino, Kaitlyn King, Grace Li, Carlos Rinaldi-Ramos

Magnetic nanoparticles are of great interest for biomedical applications due to their ease of synthesis, biocompatibility, and tunable physicochemical and magnetic properties. Magnetic Particle Imaging (MPI) is a novel molecular imaging modality relying on the nonlinear magnetization of superparamagnetic iron oxide nanoparticle (SPION) tracers in a time-varying magnetic field, which has potential for applications in blood pool imaging, magnetic hyperthermia, magnetically triggered drug release, and tracking cell therapies. Modeling of MPI physics suggests that iron oxide nanoparticles larger than 25 nm with uniform magnetic properties would achieve optimal performance, meaning high signal intensity and resolution. The challenge of synthesizing these tracers is that as size increases, the probability of defects and multiple crystal structures forming within a particle increases, impairing magnetic properties and hindering MPI performance. Previous efforts in our group have demonstrated that oxygen addition during synthesis can eliminate the magnetic dead layer often reported to explain suboptimal magnetic properties. This *in-situ* oxidation method has allowed our group to develop 21 nm tracers for MPI (RL-1) with magnetic properties similar to bulk magnetite, yielding 4x better sensitivity than commercial tracer Ferucarbotran. We hypothesize the method's limited success for larger particles (> 20 nm) is due to gas-liquid mass transfer limitations, which has been a barrier to synthesize tracers with superior sensitivity and resolution. To overcome this limitation, we are adopting use of a machined reactor with a gas entrainment impeller that bubbles the gas in the reactor's headspace into the reaction mixture. Synthesis conditions of our developed tracers (RL-1) were tested in the new reactor using various solvents. The effect of factors such as reactor surface, surfactant concentration, and precursor addition rates on nanoparticle growth kinetics, size distribution and morphology were explored. Our work shows the process of optimizing and troubleshooting a new reactor system for magnetic nanoparticle synthesis, which ultimately will lead to reproducible nanoparticles with superior magnetic properties and MPI performance.

269h A Simulation-Based Thermodynamic Model to Pre-Screen Peptide Amphiphile Micelle Vaccines

Luke Kruse, Bret Ulery, Karl Hammond

Vaccines have been one of mankind's greatest public health achievement with their rapid development having become ever more important due to the COVID-19 pandemic. While traditional strategies employing whole-killed or live-attenuated vaccines have been successful, the emergence of new subunit and nucleic-acid-based vaccines require new engineered devices to be broadly efficacious. An exciting alternative vaccine delivery technology is the peptide amphiphile micelle (PAM) system. This modular platform technology offers the potential to reduce the obstacles of *de novo* vaccine development, as incorporated immunogenic peptides and molecular adjuvants can be readily changed. Specifically, the shape and size of these PAMs is known to be directly linked to their ability to elicit adaptive immune responses. While promising, the incorporation of new peptides and/or adjuvants can alter micellar architecture, so a method to predict micelle structure *a priori* would be incredibly helpful to expediting vaccine research. With this in mind, we are developing a molecular thermodynamic model employing molecular dynamics simulations to screen bioactive peptide amphiphile (PA) candidates for their self-assembled micelle morphology and size. Specifically, the free energy of micellization is decomposed into the sum of different physical contributions, including that of the hydrophobic effect, and a new method for estimating the role of headgroup interactions is posed. To test our model, a method for defining particle shape has been developed and used to generate size distributions to compare to those derived from experimental small-angle x-ray scattering (SAXS) and TEM. Once fully developed, a computation-based approach for screening PA vaccine candidates holds tremendous potential for achieving truly "warp speed" vaccine development.

461a Influence of Impeller Geometry on the Formation of Spherical Agglomerates

Victoria Kitching, Kate Pitt, Bilal Ahmed, James D. Litster, Rachel M. Smith

Spherical agglomeration is a promising particle size enlargement technique for pharmaceutical manufacturing, with ongoing research for industrial applications. In spherical agglomeration the primary crystals are suspended in a solvent system and a bridging liquid is added to create agglomerates. Most research is conducted at bench-top scale, with small stirred tanks commonly used for spherical agglomerate production. While the effect of scale and impeller geometry is known to have strong effects on flow patterns and fluid and particle velocity profiles for suspensions in stirred tanks, little investigation has been conducted on these parameters for spherical agglomeration.

To investigate the effect of impeller configuration on spherical agglomeration, computational fluid dynamics (CFD) simulations were performed using ANSYS Fluent 19.1. Impeller geometries, clearances, and speeds were varied. Flat-blade, pitched-blade, Rushton turbine, and a propeller impellers were tested. These impellers were selected to vary the relative contributions of axial and radial flow on the velocity profile of the stirred tank. The geometry constructed for the CFD simulations was a replication of a 1L stirred tank, used for complementary spherical agglomeration experiments. Standard geometric relationships for stirred tank design were used to determine the impeller clearances; tested clearances are between 1/5 to 1/3 of the tank diameter. Each impeller geometry and clearance were simulated for impeller speeds of 300rpm, 450rpm, and 600rpm.

For the experiments, a suspension of monosized 52μm PMMA plastic beads in water was used as model suspended particles, and toluene was added as a bridging liquid.

Increasing the impeller speed resulted in more spherical and denser agglomerates as increased collision velocity increases the rate of agglomeration and consolidation. Collision velocity also increased when radial flow was promoted in the tank; such as when a Rushton turbine or flat-blade impeller was used.

Lower clearances resulted in a broad particle size distribution; containing unagglomerated primary material and large, dense agglomerates which sank to the bottom of the tank, closer to the impeller flow field. Here they contacted more bridging liquid, increasing growth. This is more apparent at lower impeller speeds as there is insufficient shear to break the larger agglomerates and the velocities further away from the impeller are not high enough to allow for the smaller particles to become entrained in the flow field. This demonstrates that the clearance has a significant impact on mixing efficiency.

The CFD results have been useful to understand the flow behaviours in the stirred tank under the various operating conditions; and how these have influenced the formation of spherical agglomerates. Even though impellers induce the same type of flow in the tank the agglomerates do not undergo the same velocity or shear stress profiles, resulting in differences in the agglomerate formation. The results of the CFD simulations will be used to inform the derivation of an agglomeration kernel that will be implemented into a population balance model (PBM) for spherical agglomeration. This is done with the intention to predict ideal impeller design to produce spherical agglomerates with desired characteristics.

461b Novel Approach for the Characterization of Powder Caking

Alessandra Hausmann, Britta Buck, Laura Shaw, Tom Simons, Daryl Williams

Storage and transport conditions such as temperature and humidity can have a wide range of effects on the physical properties of powders such as loss of pharmaceutical potency, food spoilage and powder caking. Powder caking is the undesirable process of a powder forming agglomerates, referred to as "cakes". Industrial methods of determining caking conditions are often time-consuming, require large sample volumes and provide little scientific insight. However, current analytical techniques rarely represent "real-world conditions" and may provide inaccurate data when compared to the storage and transport lifetime of the powder sample.

We have developed a novel characterization method for assessing the role of humidity in caking, allowing for in-situ measurements of powder flow while exposed to a tunable environment of humidities. It is faster than current industry standards and correlates to common methods used. The technique uses commercially available equipment such as the FT4 Powder Rheometer, correlates to common experimental methods used in solids handling industries and is easily reproducible due to its modular design. Among other parameters, the instrumental setup measures flowability energy, as a function of relative humidity. Powder flow energy is a measure for a powder's resistance to flow. Parameters that influence powder flow energy include particle size, shape, cohesivity, density, elasticity and electrostatic charge.

To further understand the interaction of relative humidity with the studied powders and the mechanism of caking, supporting DVS (Dynamic Vapor Sorption) experiments were performed. DVS is a gravimetrical analysis tool that measures a materials uptake of water vapour and potentially a variety of other gases while allowing tight control over the temperature in ranges typical for industrial storage and transport (10-50°C), (0-95% RH). We were especially interested in equilibrium sorption capacities and the kinetics of the water uptake. DVS can be combined with Raman and near IR spectroscopy to provide insight in phase transitions, swelling behaviour and deliquescence of materials.

Our measurement protocol is designed to assess caking conditions, but the flexibility of the setup has the potential to be utilized more broadly for determining the moisture-induced changes in powder rheology for example in wet granulation, drying and mixing processes. This way of characterization can be used for a wide range of powders differing in crystallinity, morphology, particle size and chemical structure.

461c Granulation of Reactive Pyrotechnic Composition for Temperature Mapping of Coal-Boilers

Swarom Kanitkar, Biswanath Dutta, Daniel Haynes, Edward Sabolsky, Benajmin Chorpening

Coal is expected to remain a viable energy source for the foreseeable future in some regions of the United States due to its abundance and low cost. However, improving the efficiency of coal-fired boilers is essential to limit the emissions per unit of energy produced. Improved knowledge of the spatial distribution of temperature inside the boiler can provide insights into the boiler performance and can also be used for the tuning and validation of detailed models of combustion within the coal-fired boilers for more efficient operation. The harsh temperatures and reactive environment within a boiler often require novel ways for the determination of a 3-D temperature map of the boiler.

Pyrotechnics are flexible materials that can be tailored to produce a distinct light emission which could therefore enable them to be used as optical sensors. The requisite thermal protection could be provided by coatings designed to shed after specific thermal history. The coated pyrotechnic materials would be fed with the pulverized coal and the light emission can be captured by a camera or spectroscopic techniques to provide the 3-D map of temperature. Given the yellow-orange-red background inside the coal-boiler, a green color would be distinguishable, and thus green light-emitting pyrotechnic compositions were developed. The composition consisted of a mixture of metallic fuel, oxidant, colorant, and binder. These components need to be homogenized into granules having similar particle size to the pulverized coal to enable similar fluidization behavior and for the application of the protective thermal coating layer.

The current work focuses on the agglomeration of optimized green light-emitting pyrotechnic composition in desired-sized granules between 500 and 850 μm (20 – 35 mesh) using a high-shear mixer. The process was optimized by varying different parameters such as inter-impeller distance, type of binder, speed of rotation, and liquid to solid ratio. The granulated material was characterized in size using an optical microscope and standard sieve meshes. It was observed that inter-impeller distance negatively influenced the yields (lower distance gave higher weight% yields) as seen from figure 1. This was because a narrow gap provided more chopping action and prevented large agglomerates from forming. When the distance was increased, larger granules were formed in between the two impellers thereby drastically reducing the yield. In terms of the type of binder, ethylcellulose-based binder provided high yields as compared to conventional pyrotechnic binders: shellac and red gum. This result might have been due to the difference in the viscosity of these binder solutions. Medium speed of rotation (500 rpm) provided adequate granule growth and provided the best yields whereas, a liquid to solid ratio of 0.25 ml/g provided high yields.

Figure 1: Effect of inter-impeller distance on the granulation yield

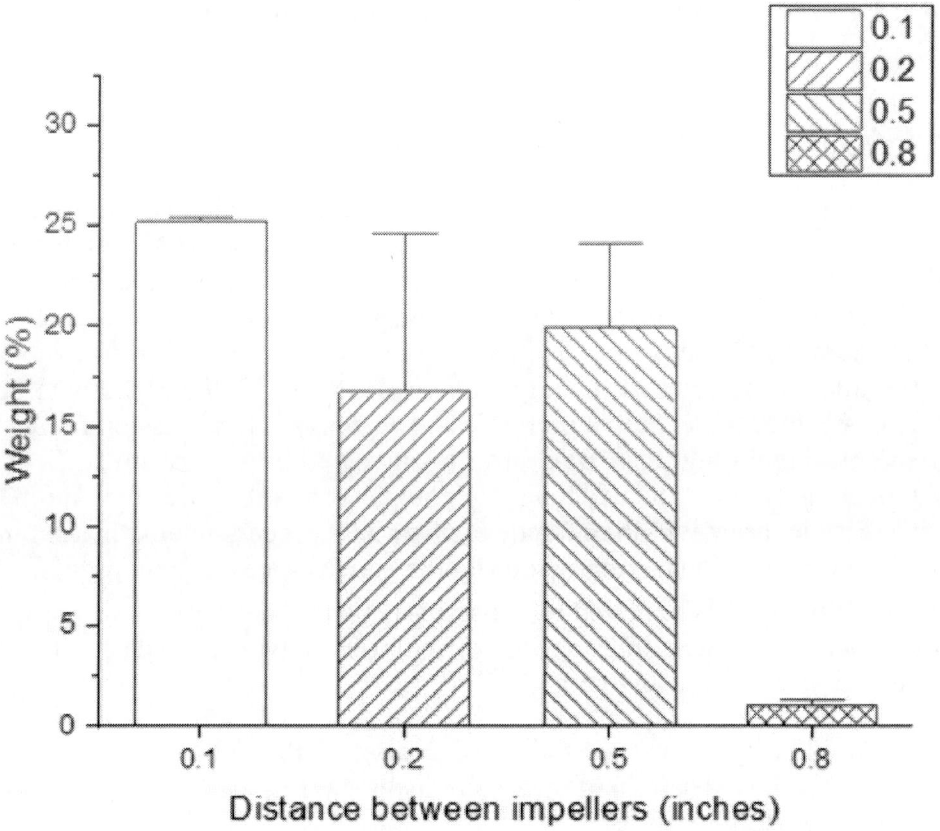
0.1
0.2
0.5
0.8
30
25
20
15
10
5
0
Weight (%)
0.1
0.2
0.5
0.8
Distance between impellers (inches)

461d Sequential Fixed Fluidized Bed Foam Granulation (SFFBFG) and Drying: Multivariate Model Development for Water Content Monitoring with Near-Infrared Spectroscopy

Abdoulah Ly, Ryan Gosselin, Esma Ines Achouri, Nicolas Abatzoglou

Foam granulation is a non-conventional technique in which the binding solution is sprayed as an aqueous foam onto a powder bed. This method could offer some advantages including a better binder distribution along with the usage of less binder compared to conventional spray granulation. In this work, the aqueous foam is sprayed onto a pharmaceutical powder blend followed by an agglomeration process in a fluidized bed granulator. The main objectives of this work are (a) to covert a lab-scale fluidized bed dryer (FBD) into a fluidized bed foam granulator (FBFG) with the capacity to be used as a FBD following the FG step; (b) to develop a partial least squares (PLS) regression model for granules water content monitoring during this FBFG process and (c) to bring forward sufficient evidence of the FBFG process feasibility and efficiency in comparison with more conventional techniques. The latter is the main target within our endeavor towards process intensification in pharmaceutical industrial production. Data gathered during the entire process are preprocessed by various techniques. Fist, baseline shift is removed by Savitsky-Golay 1st derivative. Secondly, data are aligned by linear interpolation as batch durations are different. Finally, the 3D data is unfolded to 2D according to the observation-wise rearrangement.

The calibration model was built based on 58% of the initial dataset by Random Subset Cross-Validation () and validated externally. The external validation was performed on the remaining 42 % of the initial dataset. With respect to the results:

- Only two (02) latent variables (LV) were sufficient to capture approximatively 99.6 % of the total variance contained in the initial dataset. Furthermore, it was shown that the scores of the first LV (LV1) were related to the presence of water.
- An and of 0.93 and 1.8 % w/w are obtained respectively. Besides, the validation of the model gave an and of 0.9 and 2.4 % w/w respectively showing the ability of Near Infrared (NIR) spectroscopy to satisfactorily predict the granules water content during this non-conventional technique.
- The scanning electron microscopy (SEM) pictures validated the proof of concept and confirmed that this method could be used as a granulation technique in the pharmaceutical field and be combined with FBD within the same apparatus

461e A Combined Numerical Modelling and Laboratory Investigation into the Relationship between the Power to Mass Ratio, Turbulent Eddy Dissipation and Particle Size Distribution

Justin O'Sullivan, Kate O'Dwyer

Crystallisation is of great important in the pharmaceutical, food, and fine chemical industries. In the pharmaceutical industry specifically, it is used to separate and purify intermediates and Active Pharmaceutical Ingredients (APIs). Changes to crystal morphology is well understood in terms of the chemical factors such as solubility, metastable zone control, and antisolvent addition to achieve optimisable crystals. However, crystals also change because of the hydrodynamics in a vessel. The relationship between fluid mechanics and crystal characteristics (mean crystal size, particle size distribution (PSD), morphology) are yet to be understood. In industry, the main challenge in a crystallisation process is to obtain the desired PSD of crystals. The PSD is an important factor to consider as it will impact the product quality and purity. It also affects the downstream processes such as filtration, drying, and milling. However, the hydrodynamics in a crystallisation process are not well understood. The gap in this understanding can reduce the flexibility in the production of pharmaceuticals.

Mechanical stress parameters (turbulence eddy dissipation (TED), strain rate, power draw) were estimated using computational fluid dynamics (CFD) to determine the effect on the particle size distribution (PSD) in a cooling crystallisation of benzoic acid. In CFD the distributions of the TED and strain rate were outputted, and maximum values found for different impellers (anchor impeller, pitched blade impeller, retreat curve impeller, half-moon impeller), baffled and unbaffled reactors and using two different power draws (0.15 W/kg and 0.01 W/kg). Experiments, to replicate the simulations in the laboratory, were then run.

The experimental PSD was found to be related to an increase of TED and strain rate (found in CFD simulations). The PSD increased 26 -73 % at the high-power draw power draw with the introduction of baffles. A combination of high-power draw and change in impeller type could decrease the PSD by 3-51 %. It was determined to achieve the smallest PSD the baffled half-moon impeller configuration must be used. The mechanical stress outputs (TED and strain rate) were the largest from the CFD simulations for the half-moon impeller. This suggests the increase in the mechanical stress led to increased attrition and a smaller PSD as a result. The retreat curve impeller produced the largest PSD, which was attributed to its poor mixing performance, leading to less attrition and overall poor energy transmission. A workflow from the variables (impeller, baffles and power draw) was developed to aid in the optimisation of the PSD using mechanical stress factors.

This integrated approach between mathematical models and experiments allowed for the rapid assessment of potential practical solutions on a laboratory scale unit operation before large scale pharmaceutical manufacturing runs. The TED rate, strain rate, and power draw have been deemed important characteristics that possibly cause attrition of crystals, reducing the PSD as they are increased. This offer researchers and manufacturing additional means to control the PSD of pharmaceutical products.

333a Development of Multi-Dimensional Population Balance Using a Breakage Mode Classification Kernel for the Prediction of Milled Granule Quality Attributes

Ashley Dan, Rohit Ramachandran

Milling is an important particle design unit operation used in several manufacturing industries. It is often used post-granulation to fine-tune the particle size distribution (PSD) to meet product specifications [1]. There have been numerous studies on the modeling of milling processes using population balance model (PBM) [2-4]. Previously, implemented PBM models were one-dimensional (1-D), accounting for only one intrinsic property, usually the particle size. This approach assumed no considerable variations in other characteristics within a size class and did not account for changes in other particle properties [5]. However, granule properties, such as porosity, can impact the breakage process and affect the milled granule quality attributes [6]. Therefore, there is a need to develop a higher dimensional PBM that accounts for changes in PSD, porosity, and content uniformity.

Breakage modes describe the granule breakage behavior and are dependent on the granule properties and the force applied to it. Particle breakage in a milling process such as the Comil is known to undergo impact and attrition mode. Each mode can occur individually or simultaneously during the mill operation, which is attributed to the process conditions and material properties. Materials with lower porosity undergo impact-attrition mode throughout the breakage process, whereas materials with higher porosity undergo impact mode in the beginning and transition into impact-attrition breakage mode as the mill approached steady-state [7]. Current PBMs simulate the milling process by calculating breakage kernels using only one type of breakage mode which limits their applicability. Different breakage modes within the mill can alter the final granule quality attributes. Thus, it is essential for the model to be able to incorporate different breakage modes and calculate breakage kernels based on the classification between these modes based on the process conditions and material properties.

In this study, a novel multi-component population balance model (PBM) is being developed that incorporates the classification of breakage modes into its breakage kernel. The model can predict PSD, porosity, and content uniformity. The breakage kernels in the PBM account for interactions between multiple solid components (e.g., API and Excipient) and gas volume. The kernel equations are a function of impeller speed and material properties, such as particle size and porosity. A classification kernel is being used to distinguish between the different breakage modes that the mill undergoes, giving the model the ability to simulate different CQAs produced at different breakage modes. Experimental data collected will be used to optimize the estimated parameters in the proposed kernels and validate the model results. The proposed PBM model will accurately predict an increase in the average porosity within each size class and an improved content uniformity post mill, where super-potency with respect to the API in the fines region is diminished. The model also accurately predicts a bimodal PSD for materials with lower porosity and a unimodal distribution for materials with higher porosity.

References:

1. Fassihi, A.R. and I. Kanfer, Effect of Compressibility and Powder Flow Properties on Tablet Weight Variation. Drug Development and Industrial Pharmacy, 1986. 12(11-13): p. 1947-1966.
2. Barrasso, D., et al., Population balance model validation and predictionof CQAs for Continuous milling processes: Toward QbDin pharmaceutical drug product manufacturing. Journal of

Pharmaceutical Innovation, 2013. 8(3): p. 147-162.
3. Metta, N., et al., Model development and prediction of particle size distribution, density and friability of a comilling operation in a continuous pharmaceutical manufacturing process. International journal of pharmaceutics, 2018. 549(1-2): p. 271-282.
4. Reynolds, G.K., Modelling of pharmaceutical granule size reduction in a conical screen mill. Chemical Engineering Journal, 2010. 164(2): p. 383-392.
5. Iveson, S.M., Limitations of one-dimensional population balance models of wet granulation processes. Powder Technology, 2002. 124(3): p. 219-229.
6. Mirtič, A. and G.K. Reynolds, Determination of breakage rate and breakage mode of roller compacted pharmaceutical materials. Powder Technology, 2016. 298: p. 99-105.
7. Dan, A., et al., Understanding the effects of process parameters and material properties on mill regimes and breakage mechanisms (Under preparation).

333b Dry Ball Milling of a Pre-Milled Cement Clinker: Elucidating the Impact of the Ball Material Type and Size of Grinding Balls

Ecevit Bilgili, Nontawat Muanpaopong, Rajesh Dave

A modern cement milling process usually incorporates a ball mill with a hydraulic roller press (HRP) as a pre-milling unit. The use of an HRP not only reduces the feed size of the particles before feeding to the ball mill, it also generates more flaws–defects in the particles [1]; thus, making it easier to break the particles in the mill. In the cement industry, one of the current topics for cost reduction for ball milling is to determine how to replace traditional steel grinding balls with different materials, including alumina ceramic [2]. The rationale for this replacement is that the density of a ceramic ball is less than half that of a steel ball, resulting in lower absorbed mill power. Details of a ball mill power consumption model can be found in [3], where mill power is reported to be directly proportional to the grinding ball density for a given operating condition.

Population balance modeling (PBM) is a mathematical description of various milling processes, including ball milling [4]. A PBM is based on two fundamental functions (parameters): a specific breakage rate parameter S_i and breakage distribution parameter. The direct measurement methods such as the one-size-fraction method to determine S_i [5] and the BII method to determine the cumulative breakage distribution parameter B_{ij} [6] as well as the optimization-based back-calculation method [7] have been used to quantify the PBM parameters. Obvious disadvantages of the direct measurement methods are that they require time-consuming preparation of various mono-sized feeds and the need to conduct many experiments. Additionally, based on commercially available sieve sizes, fine particle sizes cannot be experimentally prepared; thus, extrapolation to the fine particle size of interest is required. On the other hand, the back-calculation approach has been shown to be a more efficient way to determine the PBM parameters. The current state of knowledge for the back-calculation method [8] was to combine a local optimizer (with a single set of initial guesses) and an ordinary differential equation (ODE) solver to minimize the difference between the experimental data and the predicted values. With a single set of initial guesses, it is possible that the optimizer converges to a local minimum or flat surface region, which can be mitigated by a uniform wide-range initial value search [9].

To the best of our knowledge, no previous study has quantified the PBM parameters (i.e., S_i and B_{ij}) via a global optimizer-based back-calculation method for a pre-milled cement clinker milled in a tumbling ball mill and compared the performance of steel vs. ceramic grinding balls. We conducted milling experiments using a lab-scale ball mill (diameter and length of 300 mm) with a fixed rotation speed of 71 rpm. A pre-milled clinker by HRP was extracted from an actual cement plant. The controlled conditions are the ball filling fraction and the interstitial power filling fraction, kept constant as 0.3 and 0.35, respectively. Three different single size steel balls (20, 30, and 40 mm) were selected for the experiments as these ball sizes are commonly used in industrial ball milling. To validate the PBM, an additional experiment was carried out with a mixture of steel balls (20, 30, and 40 mm) based on uniform ball mass. All the experiments were repeated using a ceramic ball. Samples were taken for particle size distribution (PSD) analysis by sieve shaker and laser diffraction down to 1.8 mm at milling times of 0, 2, 4, 8, 12, 20, and 28 min. We did not attempt to mill longer than 28 min as PSD at 28 min is approximately in the range of regular cement fineness.

As a major novelty, we developed a coupled, global optimizer–ODE solver based back-calculation method with multiple trial points using MATLAB's global optimization toolbox to fit the experimental PSD evolution data. Unlike previous works [8,9], the GlobalSearch function, a scatter-search method for generating trial points as part of the MATLAB version 9.11 global optimization toolbox, along with the constrained optimization function, fmincon, were used to find the global minimum of cumulative PSDs between the experimental values and the predicted values. During each iteration of the optimizer, a size-discrete, time-continuous, well-mixed batch PBM equation was numerically solved using the ODE solver, ode15. Specifically, the number of trial points was set to 10^4. Along with the non-normalized function for

B_{ij}, we selected the Kotake–Kanda model for S_i [10] for back-calculation for the first time because of the lower number of parameters in comparison to the Austin model (see their similarity in [11]). The total number of parameters to be back-calculated was 9, broken down to 5 and 4 parameters, respectively, for S_i and B_{ij}.

Our newly developed back-calculation method fitted the experimental data reasonably well. The results are not shown for brevity here. Several insights can be drawn from the S_i and B_{ij} parameters obtained from the back-calculation: (i) A larger ball size is more effective in breaking the coarser particles, while a smaller ball size is effective in breaking finer particles, as intuitively expected; (ii) For a given ball size, a steel ball breaks particles faster than a ceramic ball for the entire range of particles; (iii) Maximum S_i does not change according to the ball size, which is consistent with previous experimental observations [10]; (iv) In view of the distribution of primary breakage, a steel ball produces slightly finer particles than a ceramic ball, which is in line with [12]. Subsequently, the PBM parameters obtained from the back-calculation method for both the steel ball and the ceramic ball were then used to predict the timewise evolution of PSD using a mixture of different ball sizes. Figure 1 shows the predictive capability of the PBM in comparison to the experimental data sets that were not used during the back-calculation, showing good agreement between the experimental and predicted evolution of the PSD.

The physical origin of the impact of the material type (steel and ceramic) and the size of the balls on particle breakage was then explained by a discrete element method (DEM) simulation. Here, only grinding balls without cement clinker particles were simulated. The input parameters, including the material properties and coefficients, were all taken from the literature. As our purpose was to gain insights from the DEM simulation regarding the microdynamic behavior of steel vs. ceramic balls, a calibration of the coefficient parameters has not been performed, which is beyond the scope of this work. For a given ball size, steel balls are associated with a higher collision frequency at the high collision energy levels based on dissipated energy ($> 1\times10^{-3}$ J), while ceramic balls are associated with a much higher collision frequency at the low collision energy levels than the steel balls. Additionally, for a given ball material, collisions of larger balls generate higher collision energy levels than those of smaller balls. These findings explain the origin of faster breakage observed with the steel vs. ceramic balls.

Overall, the PBM parameters for the pre-milled cement clinker using steel–ceramic balls, obtained from our developed back-calculation method, have conclusively demonstrated the superiority of steel balls vs. ceramic balls in terms of the breakage kinetics in a batch tumbling ball milling process. The DEM simulations provided significant insights into the microdynamic origin of the better performance of steel balls.

References

[1] I.B. Celik, M. Oner, The influence of grinding mechanism on the liberation characteristics of clinker minerals, Cem. Concr. Res. 36 (2006) 422–427.

[2] T. Nowack, Cement grinding is changing–trends towards decarbonisation, Optimising the Grinding Process, 2022 Live Webinar Series, March 2022.

[3] S. Brochot, R.L. Wiegel, S. Ersayin, S. Touze, Modeling and simulation of comminution circuits with USIM PAC, in: S.K. Kawatra (Ed.), Advances in Comminution, Society for Mining, Metallurgy, and Exploration, Inc., Colorado, 2006, pp. 495–511.

[4] E. Bilgili, B. Scarlett, Population balance modeling of non-linear effects in milling processes, Powder Technol. 153 (2005) 59–71.

[5] L.G. Austin, V.K. Bhatia, Experimental methods for grinding studies in laboratory mills, Powder Technol. 5 (1971/72) 261–266.

[6] L.G. Austin, P.T. Luckie, Methods for determination of breakage distribution parameters, Powder

Technol. 5 (1971/72) 215–222.

[7] R.R. Klimpel, L.G. Austin, The back-calculation of specific rates of breakage and non-normalized breakage distribution parameters from batch grinding data, Int. J. Miner. Process. 4 (1977) 7–32.

[8] M. Capece, E. Bilgili, R. Dave, Identification of the breakage rate and distribution parameters in a non-linear population balance model for batch milling, Powder Technol. 208 (2011) 195–204.

[9] J. Kwon, H. Cho, Investigation of error distribution in the back-calculation of breakage function model parameters via nonlinear programming, Minerals 11 (2021) 425.

[10] N. Kotake, K. Suzuki, S. Asahi, Y. Kanda, Experimental study on the grinding rate constant of solid materials in a ball mill, Powder Technol. 122 (2002) 101–108.

[11] E. Bilgili, On the similarity of Austin model and Kotake–Kanda model and implications for tumbling ball mill scale-up, KONA Powder Part. J., in press.

[12] Y. Umucu, V. Deniz, The effect of ball type in fine particles grinding on kinetic breakage parameters, Inzynieria Mineralna (2015) 197–203.

333c An Enthalpy-Balance Model for Timewise Evolution of Temperature during Wet Stirred Media Milling of Drug Suspensions

Gulenay Guner, Sherif Elashri, Mirsad Mehaj, Natasha Seetharaman, Dogacan Yilmaz, Ecevit Bilgili

In this study, milling experiments were performed at three different stirrer speeds, bead loadings, and bead sizes using 10% Fenofibrate, 8% hydroxypropyl cellulose and 0.05% sodium dodecyl sulfate (w/v) formulation. If and when the mill outlet temperature reached 45 oC, the mill was shut down and cooling was continued. If not, the milling continued for a total of 60 min and the temperatures were recorded at each minute. Chiller temperature was recorded at each time point as well. Timewise evolution of the mill outlet temperature and the chiller temperature for milling part and the cooling part were used as input files for our MATLAB code. Two dynamic enthalpy balances (ordinary differential equations in time) were developed: one for the mill content and another for the holding tank content. For the mill, we considered only a fraction of the mechanical power that is converted to heat (ξ) as well as the enthalpy of the recirculating suspension between the mill chamber and the holding tank. Both the mill and the holding tank have cooling jackets through which cooling water passes through and returns to the chiller. The heat removal rate through the jackets was calculated using the Newton's law of cooling with the respective overall heat transfer coefficient. Thermophysical properties for the mill chamber and the holding tank were found or calculated by using correlations from the literature. All physical properties were known except for the ξ, which was found by dynamic optimization of the experimental data and the model by fmincon function of MATLAB. Overall, the model could fit the experimental data well as signified by the low residual sum of squares.
While breakage kinetics during WSMM have been studied in the literature (see ref [6], and the references cited in there) using statistical and mechanistic models, another major concept, the heat dissipation, has not been experimentally studied or modelled. Heat dissipation (generation) is a crucial aspect in pharmaceutical applications that limits the selection of process conditions and restricts the process design space accordingly. Guner et al. [7] highlighted the importance of simultaneous consideration of heat dissipation besides the breakage kinetics and energy consumption. Since the drugs and polymers used in the formulation may be thermolabile, the temperature should be controlled stringently. Not all of the mechanical energy that was put into system is used for particle breakage and most of it is converted to heat causing the temperature to increase. Overheating was mentioned and prevented by applying intermitting milling–cooling cycles in some applications [8, 9], however, there is no comprehensive study that investigates the heat dissipation in WSMM using a macroscopic enthalpy balance approach.
Nanosuspensions have been established as one of the platform approaches to address the poor solubility of drugs over the last few decades [1]. Nanoparticles can improve the bioavailability of such drugs owing to their higher surface area, increased saturation solubility, and reduction in the diffusion layer compared with micron-sized particles [2]. Additionally, nanosuspensions offer a high drug dose that allows less oral unit administration and reduced toxicity and side effects [3]. Among all techniques [4, 5], wet stirred media milling (WSMM) has been the first choice for drug nanosuspension production. In WSMM, drug particle size and breakage kinetics can be easily adjusted by tuning the process parameters and/or bead properties.
The next step is making simulations and predictions of the temperature profile for any given condition. Power consumption (P) is an input in the MATLAB code, which is read from the mill screen during the experiment. Also, we found ξ by fitting the model to experimental temperatures. If one can find a way to predict P and ξ, one can predict the temperature profiles as well. Two approaches were used to predict P and ξ: power law model and machine learning approaches. P was fitted by a power law model of dimensionless process condition numbers and for ξ, after trying several power law models based on process conditions, it was found to be the most correlated with P so it was predicted based on that. On the other hand, several machine learning algorithms were tried to predict P and ξ based on the process parameters (stirrer speed, bead loading and bead size), and k-nearest neighbor with $k = 3$ provided the best predictions. Figure shows the experimental data, enthalpy balance model fit and predictions by machine learning and power law approaches of 5 experiments that were used to test the models. It can be seen while all models provide a fairly realistic simulation of the temperature profiles, machine learning approach was slightly better compared to the power law approach. To conclude, this study was the first to use an

enthalpy balance model that captures all relevant thermophysical properties of WSMM process to investigate the temperature profiles when process conditions are varied. Furthermore, it was shown that it is possible to predict a temperature profile for any given condition without performing additional experiments, which would be a great help for engineers when selecting the proper process conditions.

References

[1] P. Kocbek, S. Baumgartner, J. Kristl, Preparation and evaluation of nanosuspensions for enhancing the dissolution of poorly soluble drugs, International journal of pharmaceutics, 312 (2006) 179-186.
[2] R. Shegokar, R.H. Müller, Nanocrystals: Industrially feasible multifunctional formulation technology for poorly soluble actives, International journal of pharmaceutics, 399 (2010) 129-139.
[3] B.E. Rabinow, Nanosuspensions in drug delivery, Nature Reviews Drug Discovery, 3 (2004) 785-796.
[4] A. Bhakay, M. Rahman, R.N. Dave, E. Bilgili, Bioavailability enhancement of poorly water-soluble drugs via nanocomposites: Formulation–processing aspects and challenges, Pharmaceutics, 10 (2018) 86.
[5] J.P. Möschwitzer, Drug nanocrystals in the commercial pharmaceutical development process, International Journal of Pharmaceutics, 453 (2013) 142-156.
[6] E. Bilgili, G. Guner, Mechanistic modeling of wet stirred media milling for production of drug nanosuspensions, AAPS PharmSciTech, 22 (2020) 2.
[7] G. Guner, M. Kannan, M. Berrios, E. Bilgili, Use of bead mixtures as a novel process optimization approach to nanomilling of drug suspensions, Pharmaceutical Research, 38 (2021) 1279-1296.
[8] G. Guner, D. Yilmaz, E. Bilgili, Kinetic and microhydrodynamic modeling of fenofibrate nanosuspension production in a wet stirred media mill, Pharmaceutics, 13 (2021) 1055.
[9] M. Li, N. Yaragudi, A. Afolabi, R. Dave, E. Bilgili, Sub-100nm drug particle suspensions prepared via wet milling with low bead contamination through novel process intensification, Chemical Engineering Science, 130 (2015) 207-220.

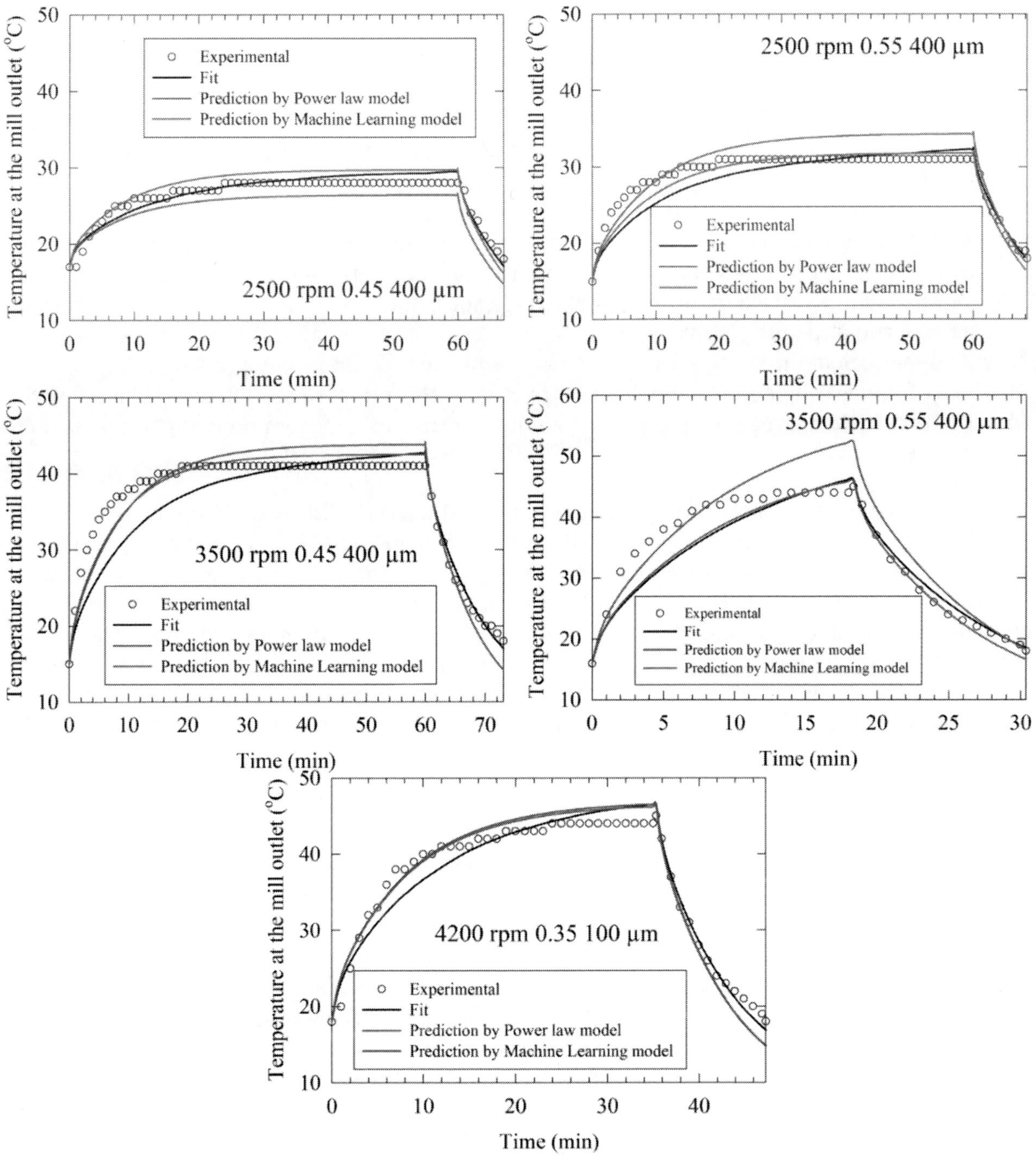

Temperature at the mill outlet (°C)
Time (min)
Experimental
Fit
Prediction by Power law model
Prediction by Machine Learning model
2500 rpm 0.45 400 μm
2500 rpm 0.55 400 μm
3500 rpm 0.45 400 μm
3500 rpm 0.55 400 μm
4200 rpm 0.35 100 μm

333d Modelling Particle Aggregation and Breakage in Suspensions By Coupled CFD-DEM

Lequan Zeng, Eirini Goudeli, George Franks

Particle aggregation and dispersion in suspensions are two important processes in the mineral industry because the particle characteristics (e.g., size structure, shape) highly influence the flow properties of suspension (e.g., viscoelasticity, particle settling rate and shear stress), which further affect downstream product grade, recovery and process efficiency. Especially in froth floatation, the physical characteristics of aggregates can determine the collection of valuable minerals. Therefore, the control of dispersion and aggregation of fine particles is significant for the mineral processes. Changing pH, adding surface modifiers, or applying shear force play an important role in controlling aggregate size and structure (Franks and Zhou, 2010). Although much research has focused on aggregate dynamics in the suspension, the mechanisms are still not fully understood due to the difficulties of monitoring aggregate behaviour over very short time steps. Therefore, simulation and modelling of particle aggregation and dispersion in suspensions is becoming a imperative to study aggregate dynamics.

In this work, two numerical models, Discrete Element Method (DEM) and Computational Fluid Dynamics (CFD), were coupled to simulate the effect of shear and pH on aggregate dynamics in suspension. This simulation describes the motion of particles and fluid, particle-particle interaction, and particle-fluid interaction. The structural properties of aggregates (cumulative size distribution, fractal dimensions, average size, etc.) were determined when the aggregate dynamics reached a stable condition in the simulation. Accordingly, a higher shear rate increases the chances of particle collision, but also leads to faster aggregate breakage, in agreement with Kushimoto et al. (2020). Aggregates with higher fractal dimension have a dense, spherical-like structure with more neighbour particles, and are much harder to break into smaller fragments.

References

Franks, G. V.; Zhou, Y., Relationship between aggregate and sediment bed properties: Influence of inter-particle adhesion. Advanced Powder Technology 2010, 21 (4), 362-373.

Kushimoto, K.; Ishihara, S.; Pinches, S.; Sesso, M. L.; Usher, S. P.; Franks, G. V.; Kano, J., Development of a method for determining the maximum van der Waals force to analyze dispersion and aggregation of particles in a suspension. Advanced Powder Technology 2020, 31 (6), 2267-2275.

333e Effect of Agitation Rate on Breakage Kinetics of Urea Crystals in Agitated Slurries

Paras Shah, Priscilla Hill

Particle breakage can occur in stirred vessels such as solution crystallizers and often occurs when the crystals are acicular, or have high aspect ratios. These acicular crystal are frequently produced when crystallizing pharmaceuticals or fine chemicals. Crystal breakage is important because it can produce many fines that can act as secondary nuclei, or that can plug the filter downstream of the crystallizer. Producing secondary nuclei may or may not be desired, but plugging the filter is not desired. To better operate crystallizers, it is necessary to be able to model the breakage of high aspect ratio particles. Therefore, this work investigates the breakage of high aspect ratio particles at various agitation rates.

Previous researchers [1-2] modeled the specific rate of breakage of high aspect ratio crystals as being proportional to the cube of the agitation rate. In one case experiments were only performed at one agitation rate [1], while there were not experimental comparisons for the other case [2]. Previous researchers used simplified models where each parent particle breaks into two child particles. In these cases the parent particles were modeled as rectangular parallelepipeds where the width ad depth were the same and the width was the minor axis, while the length was the major axis. Some researchers modeled the breakage as occurring perpendicular to the long axis of each crystal, while others modeled breakage as occurring parallel to the long axis of each crystal. It was also determined that there is probably both types of breakage occurring.

This investigation used urea as the experimental compound. Urea crystals were grown in water to produce high aspect ratio crystals for the breakage experiments. Crystals were sieved to narrow the size distribution used for the breakage experiments and to reduce fine particles. Before starting the breakage experiments, the initial parent crystal were characterized to determine the length and width of each crystal so that the aspect ratio of each crystal could be determined.

Breakage experiments were conducted in an agitated vessel with the urea crystals suspended in hexane. Hexane was used because urea is practically insoluble in hexane, and this isolates the mechanism of breakage by preventing growth and dissolution. Initial experiments were conducted in a lab scale vessel at 1250 rpm and crystals were characterized after 1, 5, 10 and 15 minutes of breakage. The agitation rate was chosen to match the impeller tip speed in industrial crystallizers.

Results from this research are reported as the average aspect ratio as a function of crystal length for various size intervals. Average aspect ratios were studied both as a function of the major axis and as a function of the minor axis. Our results show that the residence time for breakage has little effect on the average aspect ratio for each size interval, and that the aspect ratio appears to be a stronger function of child particle size. A comparison of the models with experimental results show that: 1) the simplified model with breakage perpendicular to the major axis cannot explain the aspect ratio distribution results, 2) the simplified model with breakage along the minor axis cannot explain the results, and 3) that a combination of the two models cannot explain the resulting aspect ratio distributions.

[1] Kazuhiro Sato, Hidetada Nagai, Kazuhiro Hasegawa, Kunihiko Tomori, H.J.M. Kramer, P.J. Jansens, Two-dimensional population balance model with breakage of high aspect ratio crystals for batch crystallization, *Chemical Engineering Science*, 63 (2008) 3271 – 3278.

[2] Botund Szilagyi and Bela G. Lakatos, Model-based analysis of stirred cooling crystallizer of high aspect ratio crystals with linear and nonlinear breakage, *Computers andamp; Chemical Engineering*, 98 (2017) 180-196.

333f A Viscoelastic Model for Droplet Breakup in Dense Emulsions

Joseph Peterson, Vipin Michael, Ioannis Bagkeris

In many models of emulsion breakup, unsteady deformations pose a problem for breakage kernels: the droplet's shape and proclivity to breakup (as well as the resultant daughter droplet distribution) depends not on the instantaneous strain rate but rather on the whole strain history. In our work, we account for this history-dependence via a viscoelastic constitutive equation for droplet shape, with strict couplings to population balance dynamics. We describe the typical shape of a droplet at different size intervals, and show how this shape changes with (1) deformation, (2) relaxation, and (3) the influx/outflux of droplets due to breakage. The last of these lets daughter droplets retain their deformation at the moment of breakup, such that polydisperse daughter droplet distributions can be realized via a cascade of binary breakage events. This partially alleviates the pressure to develop complex history-dependent daughter droplet distributions. The talk will cover (1) an overview of the model's overall structure, (2) key assumptions and approximations, and (3) predictions for both step deformations and steady shear.

333g Polydisperse Daughter Droplet Distributions from a Binary Breakage Kernel: A Viscoelastic Model for Dense Emulsions

Joseph Peterson, Vipin Michael, Ioannis Bagkeris

In many models of emulsion breakup, unsteady deformations pose a problem for breakage kernels: the droplet's shape and proclivity to breakup (as well as the resultant daughter droplet distribution) depends not on the instantaneous strain rate but rather on the whole strain history. In this talk, we show that if this history-dependence is mediated through a viscoelastic constitutive equation for the droplet shape, then a binary breakage rule is sufficient to produce complex strain-dependent daughter droplet distributions following a step-shear deformation. This result is possible because we enforce stress continuity across the breakup process (i.e. capillary stresses from daughter droplets and parent droplets match), so the entire breakage process is interpolated as a cascade of simple binary breakage events rather than a single complex breakage event. The talk will cover (1) an overview of the model's overall structure, (2) key assumptions and approximations, and (3) predictions for both step deformations and steady shear.

512a CO_2 Utilization and Storage via Carbon Mineralization and Enhanced Extraction and Separation of Energy-relevant Materials from Unconventional Resources

Ah-Hyung Alissa Park

With the increase of the global population and the growth of the world economy and industrial sector, global energy consumption has also been increasing. Considering that the fossil fuels will still be one of the major energy sources for the foreseeable future, measures need to be taken to control the atmospheric CO_2 concentration. Carbon capture, utilization and storage (CCUS) technologies are one of the approaches to decarbonize the power and industrial sectors. Among various options, mineral carbonation, which mimics the natural weathering of silicate minerals, has a potential at scale relevant to climate change mitigation. As CO_2 reacts with silicate minerals, carbon is stabilized in the form of insoluble solid carbonates for permanent carbon storage or CO_2 utilization with permanence. If this reaction is carried out in an ex-situ reactor system, solid carbonates, high surface area silica and other minor components such as iron oxide can be produced with tailored properties and separated as value-added products. In addition to natural minerals and mine tailings, alkaline industrial wastes such as iron and steel slags can also be used as feedstock for carbon mineralization. Most of the industries producing alkaline solid wastes (e.g., steelmaking, cement and aluminum plants) are also point sources of anthropogenic CO_2. Thus, carbon mineralization using their own solid waste streams and CO_2 leads to multifaceted environmental benefits including CCUS and solid waste management. Furthermore, industrial wastes often contain other valuable components such as rare earth elements. The challenge is that silicate minerals and alkaline solid wastes are chemically complex and their dissolution kinetics are very slow. In order to address these challenges and opportunities, we have focused on the fundamental understanding of dissolution and carbonation behaviors of alkaline silicate materials and integration of step-wise separations of rare earth elements from these unconventional resources.

512b Particle interactions and transport problems at different scales: From dry powder inhalation to flows in fluidized beds

Sankaran Sundaresan

Flows of particulate material mediated by interstitial gas arise in nature and in many processes. They manifest inhomogeneities that can be traced to hydrodynamic instabilities and interparticle forces that arise through contact, van der Waals, electrostatic interactions, and capillary and viscous forces associated with liquid bridges between particles. Through three examples, this presentation will pose a few unresolved questions about physics at different scales. The first example deals with dry powder inhalation, where the inhaled air drives the fluidization of carrier particles decorated with active pharmaceutical ingredient (API) particles, the release of the API from the carrier particles, and transport through the mouth-throat region. The second example deals with coarse-grid simulations of gas-particle flows, where puzzling questions about the proper model form for the drag correction for non-cohesive and cohesive particles remain unresolved. The third example deals with the contact charging of particles in fluidized beds, where the observed effects of particle size and humidity on the extent of charging leave us scratching our heads.

512c Size and Density Segregation in Dense Granular Flows

Richard Lueptow

In dense granular flows of particles differing in size, small particles fall between large particles to segregate in lower portions of the flow thereby displacing large particles upward, a process known as "percolation." Similarly, light particles rise above heavy particles as they flow due to "buoyancy." The resulting segregation due to percolation or buoyancy depends on the differences between the particles and the flow conditions. We address granular segregation in two ways. The first approach is a continuum segregation model based on the advection-diffusion equation with a term added to account for particle segregation. This model can accurately predict mixing and segregation in both steady and transient flows for a variety of flow geometries and for a range of particle systems including multiple individual particle sizes, polydisperse (continuous) particle size distributions, mixtures of particles varying in both size and density, and non-spherical particles. The model can even be used to "design" combinations of particle size, density, and concentration that result in non-segregating bidisperse particle mixtures. The second approach is to characterize the particle level segregation force via discrete element method simulations, which provides the ability to predict whether an intruder particle will rise or sink. These single intruder results can be extended to cooperative segregation phenomena in particle mixtures. Our current challenge is to connect segregation parameters in the continuum segregation model to particle level forces as well as to extend the segregation model to particle systems with a wide range of particle sizes. Funded by The Dow Chemical Company, the Procter andamp; Gamble Company, and the National Science Foundation (CMMI-1435065, CBET-1511450, CBET-1929265).

137a Demonstration of Multi-Level Coarse-Grain (MCG) DEM Simulation Technique for Twin Screw Feeder

Tarun De, Lokeshwar Mahto, Jayanta Chakraborty, Jitendra Kumar, Anurag Tripathi, William Ketterhagen, Maitraye Sen

Simulation of large granular systems using the discrete element method (DEM) is challenging because it takes a great amount of computation. One of the ways to reduce the computation expense is to use massive parallelization using Graphical Processing Units (GPUs). However, the recently published Multilevel coarse-graining technique (MCG) [1] may also be used to achieve significant speed-up. The computational time is not only dependent on the number of particles, but also on the complexity of the system geometry. Hence, additional and significant speed-up can be achieved using appropriate partitioning of the geometry among the processors and optimized mesh generation. In this presentation, we will demonstrate these strategies using the twin-screw feeder [2] shown in Figure 1(A).

In MCG simulations, particles with different coarse-grain ratios are used in different regions in the system to reduce the computational cost while preserving simulation accuracy. An example of an MCG approach with three coarse-grain ratios is shown in Figure 1(B). The MCG simulation of the twin-screw feeder has been conducted in LIGGGHTS-Public using 24 cores. Using this strategy and optimized mesh element and processor mapping, the simulation time is 4 weeks faster than the reported using GPU-based computation [2].

[1] T. De, J. Chakraborty, J. Kumar, A. Tripathi, M. Sen, W. Ketterhagen, A particle location Based multi-level coarse-graining technique for discrete element method (dem) simulation, Powder Technology (2021) 117058

[2] P. Toson, J.G. Khinast, Particle-level residence time data in a twin-screw feeder, Data Br. 27 (2019) 104672. https://doi.org/10.1016/j.dib.2019.104672.

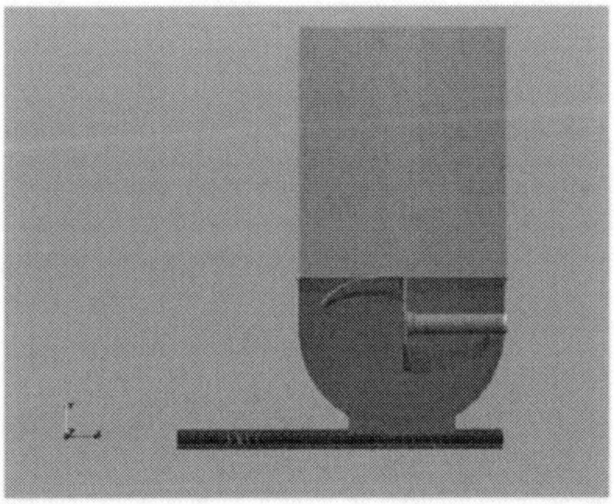

Figure 1(A): The twin-screw feeder system Figure 1(B): Snapshot during MCG simulation.

137b CFD-DEM Simulations of Pinning for Gas-Solid Flows

Kanjakha Pal, Yi Fan, Jorg Theuerkauf, Jonathan Lunn, John Holderness

The phenomenon of particle pinning occurs when crossflow of gas occurs through a moving bed of solids under gravity. Crossflow of gas occurs when gas flows orthogonal to the solids flow, such as in an annular moving bed reactor with radial gas flow. The drag force imposed by the gas flow on the solids increases the frictional force between the solids and the wall downstream of the gas flow. Above a critical value of the gas flow rate, this frictional force becomes larger than the weight of the solids bed, which causes cessation of solids flow locally near the wall, namely pinning.

In this work, we used coupled CFD-DEM method to study pinning of a moving bed of solids in a rectangular box with crossflow of gas. The CFD-DEM simulations captured the onset of pinning when gas flow is increased over a critical value. Computational investigations were carried out on how various particle and gas properties including particle size and density, gas viscosity, and particle and wall friction coefficients affect the critical velocity for onset of pinning. These simulations can form the groundwork for further studies of pinning onset in industrial reactors with gas-solid crossflow.

137c Enhanced Flowability of Dense Bi-Disperse Granular Avalanche: The Role of Fine Grains

Chongqiang Zhu, Ilaria Rucco, Fabio Dioguardi, Raffaella Ocone

Dense granular flow on inclined surfaces is encountered in engineering applications as well as geophysical flows (e.g. pyroclastic flows, landslides and avalanches) [1]. Understanding the flowability of dense granular flow is paramount for both industrial optimisation and mitigation and prevention of geo-hazards. Polydispersity is an inherent attribute of both natural and manmade granular matters. Many studies of dense granular flow deals with monodisperse or slightly polydisperse granular media [2,3], despite granular polydispersity has been proved to be a critical factor in dense granular flow behaviour [4]. In this work, we study the flowability of dense bi-disperse granular avalanches using 3D Discrete Element Method (DEM). The objectives are to reveal the flowing features of the granular system and explore the physical mechanisms underlying the observed physical behaviour.

To pursue the objectives of this work, a flume model, which is composed of two main parts, namely an acceleration zone (inclined plane) and deposition zone (horizontal plane), is employed. Two kinds of spherical cohesionless particles, differing only in size (size ratio of 2.5), are used. We keep the total mass of particles constant and vary the mass proportion of large particles. The initial granular assembly is generated by randomly pouring bi-disperse particles into reservoir under gravity to preserve a well-mixed initial state. The contact forces between particles are simulated using a linear spring-dashpot model. Both sliding friction and rolling friction are considered. The coefficient of restitution for all types of collisions is kept the same by appropriately selecting the damping coefficient. The DEM simulations are conducted using the open-source code LAMMPS (Largescale Atomic/Molecular Massively Parallel Simulator) [5].

The numerical scheme is first verified and validated using inclined chute flow by comparing with theoretical results. The flowing properties including final runout, deposition geometry, the evolution of energy and microstructures are presented and compared for cases with different mass ratio of large particles. Similar flowability, or runout, is observed for monodisperse granular assemblies, which is independent of granular size. However, the bi-disperse granular avalanches present enhanced flowability and maximum flowability is observed at a critical mass proportion of large particles. Moreover, reverse grading, in which small granular are found at the bottom and at the rear parts of final deposition, is captured. Despite the rotational energy is small in comparison with translational energy, the rotation speed of small particle is greatly increased in dense bi-disperse granular flow.

To explore possible mechanisms responsible for such a feature, we first studied the jamming volume fraction of bi-disperse particles. The rheological properties of particles are greatly related to the distance to the jamming volume fraction. The variation of jamming volume fraction presents similar trend with the runout of avalanches. Moreover, inertial number is employed to quantify mobility of granular flows in simple shear tests, which directly presents the enhanced mobility in binary particles.

The interactions between large and small granular in term of sliding and rolling friction coefficient is further investigated. We found that the rolling friction between large and small particles plays a critical role in dense bi-disperse granular flow. Specifically, the rotation of small particles can efficiently lubricate the bi-disperse granular assembly to increase its flowability. Although more simulations will be needed using large polydisperse granular media, this work provides very useful insights into understanding the effects of polydispersity on granular flowability.

Reference

[1] Goujon C, Dalloz-Dubrujeaud B, Thomas N. Bidisperse granular avalanches on inclined planes: A rich variety of behaviors. The European Physical Journal E. 2007;23(2):199-215.

[2] Chialvo S, Sun J, Sundaresan S. Bridging the rheology of granular flows in three regimes. Physical review E. 2012;85(2):021305.

[3] Zhu C, Huang Y, Sun J. Solid-like and liquid-like granular flows on inclined surfaces under vibration–Implications for earthquake-induced landslides. Computers and Geotechnics. 2020; 123:103598.

[4] Remy B, Khinast JG, Glasser BJ. Polydisperse granular flows in a bladed mixer: experiments and simulations of cohesionless spheres. Chemical Engineering Science. 2011;66(9):1811-1824.

[5] Plimpton S. Fast parallel algorithms for short-range molecular dynamics. Journal of Computational Physics. 1995;117(1):1–19.

137e Modelling Pneumatic Hopper Discharge of a Fine Bulk Solid Using a Coupled CFD-DEM Approach

Stefan Pantaleev, Ravi Teja Chaganti, Dongil Chang, Hjalte Trnka, Tue Hansen, Thomas Kvistgaard Vilhelmsen, Ioannis Fragkopoulos

Pneumatic conveying is commonly employed for the transportation of bulk solids in industrial manufacturing processes due to its safety, versatility and ease of automation. However, challenges such as plug flow, agglomeration and blockages are present in pneumatic conveying systems and lead to a lower degree of control, higher maintenance requirements and reduced product quality. These problems are particularly common for fine bulk solids and can sometimes be traced to hopper feeders of the pneumatic line where arching and ratholing result in intermittent discharge and pressure surges [1]. Developing an understanding of the complex interaction between operational parameters and bulk solid properties that gives rise to these phenomena can result in improved process operation. However, relying on physical experiments alone for this purpose may be time consuming and expensive making numerical modelling a useful component of the study that can reduce costs and provide additional insight.

Numerical modelling and analysis of a plant scale pneumatic hopper feeder system is carried out in this work, with the aim of relating the system performance to the pressure drop and the powder flowability. The system consists of an axisymmetric hopper and a horizontal pipe with an air valve downstream of the hopper as shown in Figure 1. A two-way coupled CFD-DEM modelling approach is adopted, whereby the bulk solid flow is resolved in Altair EDEM and the air flow is resolved in Altair AcuSolve.

The bulk solid is modelled via a meso-scopic approach, in which DEM particles of an intermediate scale between the physical particle scale and the scale of the system of interest are used to achieve practical computational times. The effect of particle shape is accounted for by the paired particle approach and the Edinburgh Elastic-Plastic-Adhesive contact model is used to model the visco-elastic-plastic-cohesive behaviour of the powder [2]. The bulk solid model parameters are calibrated against shear cell measurements [3] of the internal and wall yield loci of micro-crystalline cellulose.

The Euler-Lagrange methodology is used for the fluid and solid phase coupling in which the fluid domain is discretized using an unstructured tetrahedral mesh and the time-accurate flow solution is computed using Altair AcuSolve, a Galerkin Least Squares (GLS) finite element method based CFD solver. An unresolved CFD-DEM bi-directional coupling approach is used for considering the effect of the motion of the particles on the fluid flow. In this approach, the volume averaged Navier-Stokes equation is used for the fluid phase through the local averaging of fluid-particle interaction force and the solid phase volume fraction. The Gidaspow correlation is used for calculating the drag force on the particles.

The predictive accuracy of the numerical model is validated against experimental measurements of the mass flow rate of micro-crystalline cellulose in full-scale physical tests. Validation cases with a fully closed and fully opened air valve are considered in order to cover the pressure drop extremes in the system. For the closed valve case, the false-air opening is blocked completely and the pressure at the outlet is varied while in the open valve case, a constant air flow rate, obtained through experimental measurements, is applied at the air valve inlet. A comparison of the simulation and experimental results is shown in Table 1 and the computed air velocity and particle volume fraction of a closed valve and open valve case are shown in Figure 2 and 3 respectively. A good quantitative agreement is observed between the predicted and measured mass flow rates for both cases, while the corresponding simulated bulk solids flow patterns qualitatively agree with experimental observations. Taken together these observations are indicative of the validity of the modelling methodology.

The numerical model of the closed valve case was used to perform a parametric analysis with the aim of relating the applied pressure drop and bulk solids flowability to process performance. To this end, the discharge of bulk solids of two flowability extremes was simulated at a wide range of pressure drops. The predicted evolution of the mass flow rate of the low cohesion material at varying pressure drops is shown in Figure 4 and the effect of cohesion on the discharge at a fixed pressure drop is shown in Figure 5.

The results reveal a transitional flow behaviour characterised by a shift from a continuous flow to the intermittent arching behaviour shown in Figure 6. This is as a result of a complex interaction between outlet pressure and powder flowability, whereby the propensity for arching increases with the pressure drop due to an increase in the consolidating stresses in the powder bed and a corresponding increase in bulk material strength in the region of the arch.

The intermittent arching results in a non-linear relationship between the mass flow rate and pressure drop with the mass flow rate increasing sub-linearly with pressure drop. The arch stability increases with bulk cohesion, resulting in a highly discontinuous flow behaviour for cohesive materials even when discharging under low pressure drops. A significant localised increase in the consolidating stresses in the region of the outlet can be observed downstream of an arch collapse as shown in Figure 6 and this may result in unwanted agglomeration and material build-up. This is consistent with experimental observations of material build up in the region of the pipe elbow.

The results suggest that simply increasing the pressure drop may not be a remedy for poor flow in a pneumatic hopper feeder system and the relationship between the pressure drop, consolidating stress and powder flowability needs to be carefully considered. Numerical modelling work is ongoing to further understand this complex relationship with the aim of informing the optimal operation of the hopper feeder system.

References

[1] Konrad, K. (1986). Dense-phase pneumatic conveying: A review. *Powder Technology*, *49*(1), 1–35.

[2] Thakur, S. C., Morrissey, J. P., Sun, J., Chen, J. F., andamp; Ooi, J. Y. (2014). Micromechanical analysis of cohesive granular materials using the discrete element method with an adhesive elasto-plastic contact model. *Granular Matter*, *16*(3), 383–400.

[3] ASTM Standard. (2008). Standard shear test method for bulk solids using the Schulze ring shear tester. *ASTM International, Www. Astm. Org*, *i*, 1–27.

Table 1 Comparison of the transferred powder mass between the simulation and experiment.

Valve configuration	Pressure drop (kPa)	Transferred mass (in 3 sec)	
		Experiment (kg)	AS-EDEM (kg)
Closed	20	2.126	2.538
Closed	22	3.303	2.727
Closed	23	3.088	2.645
Open	2	0.466	0.375

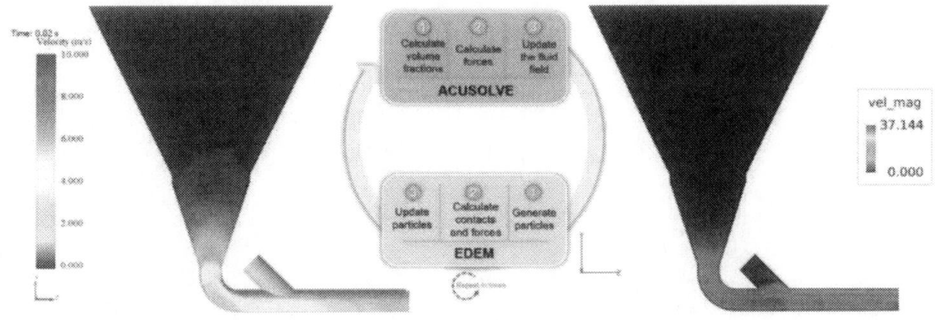

Figure 1 A two-way coupled CFD-DEM model of a pneumatic hopper feeder system.

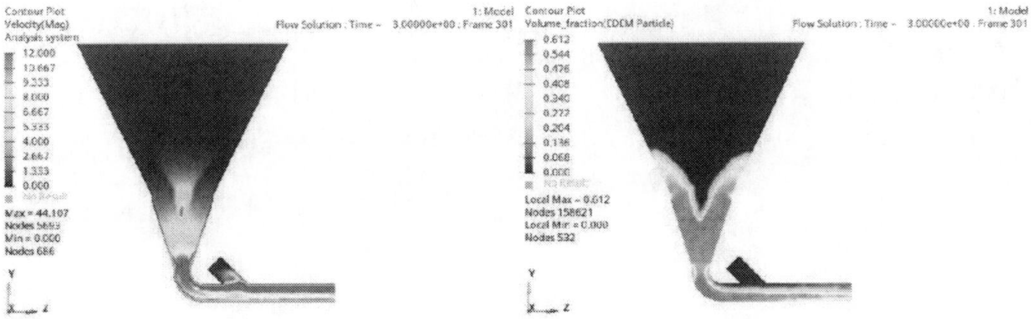

Figure 2 Velocity magnitude and solid volume fraction contour plot for closed valve case at 23kPa pressure drop.

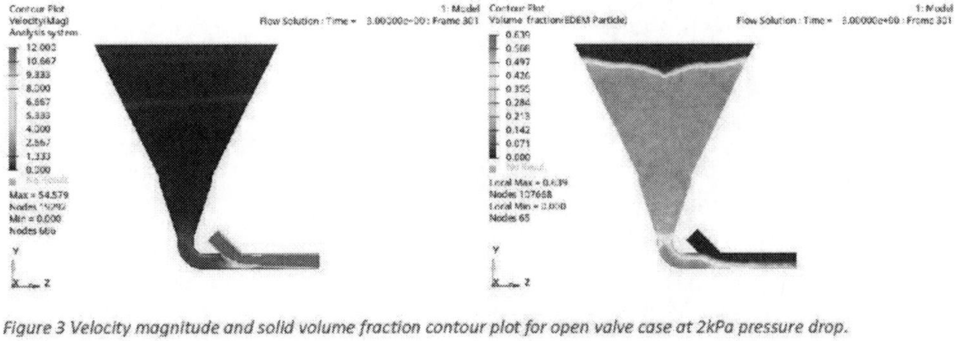

Figure 3 Velocity magnitude and solid volume fraction contour plot for open valve case at 2kPa pressure drop.

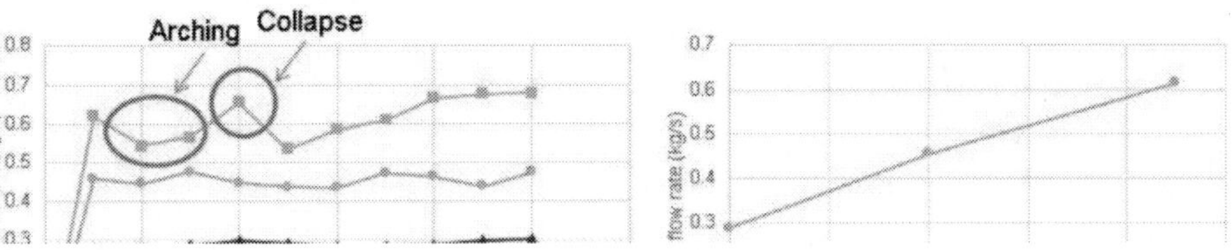

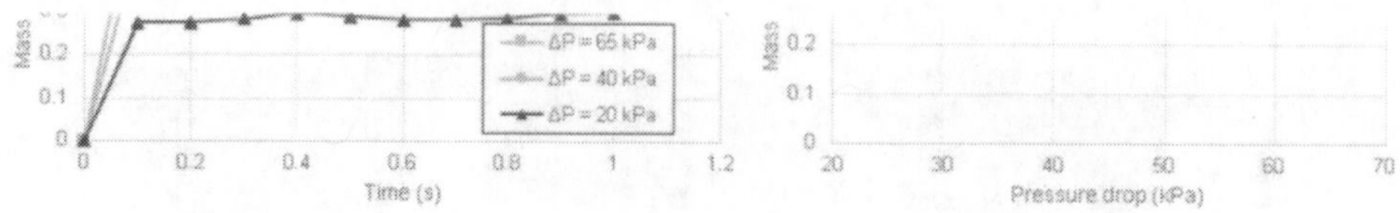

Figure 4 Predicted mass flow rate at varying pressure drops for a material of low cohesion.

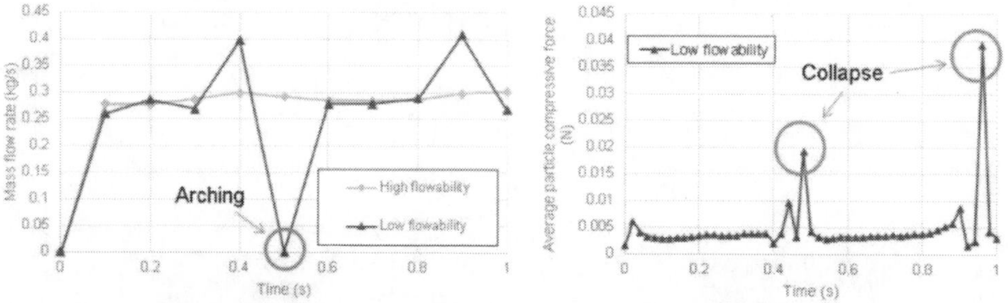

Figure 5 Predicted mass flow rate at 20kPa pressure drop for materials of two cohesion extremes.

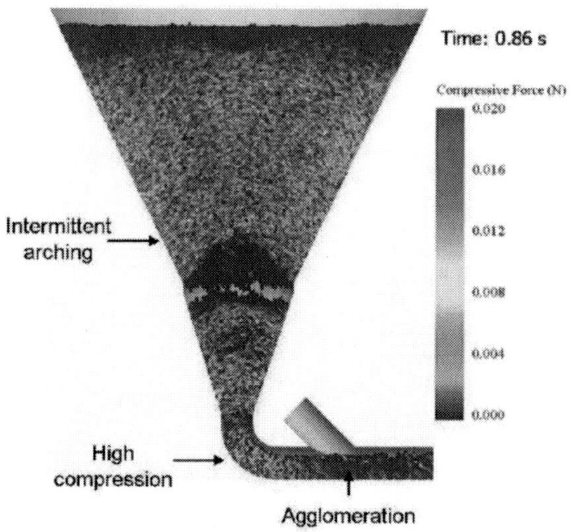

Figure 6 Intermittent arching behaviour during discharge of a low flowability bulk solid.

137f Investigation of Particle Characteristics Influence of Crumbler® Rotary Shear Comminuted Granular Biomass on the Performance of Screw Feeding: Modeling and Experiment

Ahmed Hamed, Yidong Xia, Nepu Saha, Jordan Klinger, David Lanning, Jim Dooley

Advances in bioenergy research will allow this energy source to play a vital role in shaping the clean energy portfolio to meet increasing global energy demands while satisfying the proposed measures to mitigate climate changes. Promoting fundamental understanding of the impact of the critical material attributes on the biomass feedstocks response can guide the selection of the critical processing parameters. For lignocellulosic biomass feedstocks, feeding and handling represent a substantial challenge in the supply system, due to the unfavorable properties of the raw biomass materials, e.g., low bulk density, high cohesiveness, variability, and characteristic flow behavior that involves complex localized elastic and plastic deformation patterns with strong dependency on the stress history. This poor flowability usually cause interruptions in the feeding systems and process upsets (such as jamming and clogging), which results in operation under an average feeding rate much lower than the design capacity and increased downtime. This, in turn, eventually leads to missed business opportunities and higher operation cost. Thus, identifying the underlying mechanisms controlling the mechanical behavior of this biomass feedstocks and determining critical parameters is crucial for the optimization of preprocessing unit operation and engineering favorable flowability of biomass in feeding and handling. This progress will render bioenergy technologies economically competitive and enable their deployment at commercial scale. Motivated by the quality-by-design approach, the current work investigates the effect of particle characteristics of rotary sheared pine particles on the performance of a screw feeding system using computational and experimental approaches. The system consists of a hopper, an agitator, and a screw conveyor, with materials fed at constant rpm. Using a Forest Concepts Crumbler® rotary shear system, two different particle sizes (2 mm and 6 mm) with different size distributions were produced from "matched" bulk feedstock samples from conifer veneer. Those produced Crumbles have a cubic shape. The specific energy consumption for the comminution of these Crumbles and moisture content for different particle sizes were also measured. This data can be used to inform a prospective technoeconomic analysis. To experimentally characterize the impact of particle size on the flowability of this biomass feedstock, several samples were prepared by blending the original two Crumbles sizes at different fractions and fed into the screw feeding system. The influence of particle size on the typical design parameters, i.e., the driving torque, mass flow, and energy consumption, were analyzed for different rotational velocities. Furthermore, numerical experiments were utilized to simulate the conducted experiment and elucidate the underlying mechanisms. For this purpose, discrete element method (DEM) was employed using similar feeding system and clumped-sphere model. The DEM simulation results aided in gaining valuable insights by assessing the effect of additional factors on the flowability of the Crumbles, e.g., surface roughness, cohesion model, and rolling resistance. Our experimental results showed that while smaller particle size (2 mm) of pine particles achieves better flowability (with a smaller driving torque), the energy cost of comminution is significantly higher, and the bulk density is almost the same as the 6-mm pine particles. Furthermore, the DEM results identified the mechanical interlocking between contacting particles as the dominant mechanism in determining the driving torque. Alternating flow patterns were observed in the hopper, which led to remarkable variation in the mass flow rate over the course of simulation. Simulation and experimental results were found in a good agreement. It is worth mentioning that using simple particle shapes (e.g., mono-sphere) in conjunction with an elaborate contact model to capture the plastic deformation behavior (along with rolling resistance) did not achieve the same level of agreement that was accomplished by invoking clumped-sphere particle model with a layout mimicking the shape of the corresponding Crumble, which evidences the importance of capturing the particle shape and surface roughness to reliably simulate biomass granular flow.

137g Discrete Element Modeling of Wedge Hopper Discharge of Loblolly Pine Chips

Qiushi Chen, Feiyang Chen, Yidong Xia

Forestry residues such as loblolly pine chips are a low-cost source of lignocellulosic biomass feedstocks for conversion into biofuels. Achieving efficient material handling operations (transport, feeding, etc.) of biomass feedstocks has been a challenge in the design and operation of a biorefinery. The efficiency of material handling is strongly influenced by the *flowability* of feedstocks: the propensity of a material to flow in proportion to applied stress. For Newtonian fluids, the flowability is in proportion to the inverse of the fluid viscosity. For biomass feedstocks, however, the flowability cannot be represented by a single material property like viscosity, but rather, by a collection of particle-scale attributes such as particle size distributions, shapes, density, moisture content, friction coefficient, and particle stiffness. Those attributes give rise to the bulk properties such as bulk stiffness and compressibility, internal angle of friction, cohesion, and unconfined yield stress, which collectively determine the flowability of biomass feedstocks. Feedstocks with poor flowability are more likely to cause process upsets such as clogging and jamming in material handling operations, resulting in increased downtime and higher operational cost, and consequently, lower economic value of biofuels.

In this work, we adopt a set of experiment-validated hysteretic nonlinear contact models for the discrete element method (DEM), which can capture the bulk behavior of flexible biomass particles. DEM-based particulate flow simulations of wedge hopper discharge of loblolly pine chips are performed. The DEM simulation-based study enables detailed and quantitative analyses of the hopper flow and clogging behavior based on fundamental particle physics. Wide ranges of critical processing parameters (CPPs) (of the hopper) and the critical material attributes (CMAs) (of milled loblolly pine) are considered in the analysis. Among various CPPs, hopper opening width and semi-inclination angle are the two key parameters that significantly influence the critical quality attributes (CQAs) of the hopper (i.e., discharge rate and continuity). The critical arching distance as a function of semi-inclination is predicated by DEM and compares well with experimental data and finite element method-based predictions. DEM results also show the interparticle motion resistance (a CMA) and the packing height (a CPP) have a strong influence on the critical arching distance. Findings from this work have practical implications for engineers to design and operate equipment involving biomass particle flow in biorefineries.

Experiment setup

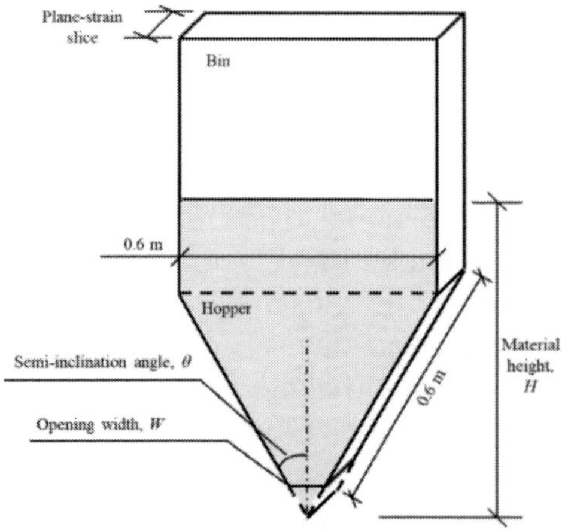

Model dimensions

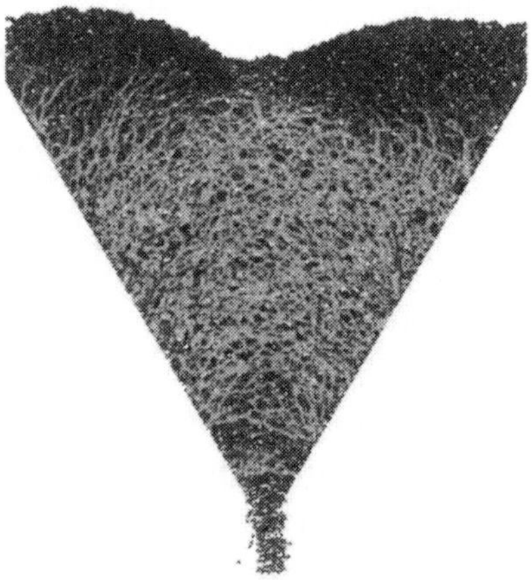

DEM simulation

137h The Elastoplastic Flexural Behavior of Corn Stalks: An Experiment-Informed DEM Model and Process Simulation of a Knife Mill

Yidong Xia, Jordan Klinger, Tiasha Bhattacharjee, John E. Aston, Vicki Thompson

This work presents the experimental characterization of the biomechanical properties of size-controlled corn stalks through a three-point cyclic bending test and the adaptation of the obtained properties to a DEM bendable particle model. From the experimental characterization, the overall elastic bending limit decreased (from 3.4 to 1.7°) with increasing stalk size, while the effective bending stiffness increased (from 7.0 to 43 $N \cdot m \cdot rad^{-1}$ in median value). The elastoplastic behavior of the stalks was more consistent and independent of stalk size where in median value the elastoplastic bending limit ranged from 6.9 to 8.8° and the ratio of elastoplastic stiffness to elastic stiffness ranged from 0.40 to 0.42. Statistical analysis of the samples and empirical results presented insightful expressions to relate the stalk scales to responses, or distributions in properties for a population of corn stalks (https://doi.org/10.1016/j.biosystemseng.2022.02.016). Parameterization of the DEM bendable particle model will be based on the experimental data of the elastic bending angle limits, elastoplastic bending angle limits, elastic bending stiffnesses, and plastic ratios. Comminution simulations of a large-scale knife mill will be conducted, where individual corn stalks will be approximated based on the parameterized DEM model. A particle breakage model in DEM will be further calibrated based on the produced particle size distribution (PSD) data from the corresponding physical comminution test of the knife mill.

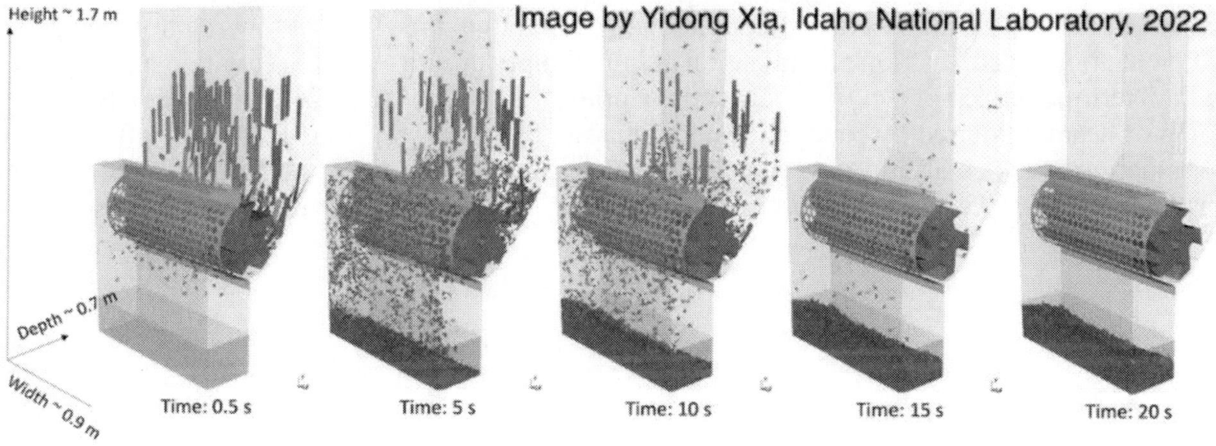

80a Effect of Low Rotation Rate on Conduction and Radiation Heat Transfer in a Rotary Drum

Bhaumik Bheda, Heather Emady

Rotary drums deliver high heat and mass transfer rates, making them extensively useful for powder mixing and heating processes in metallurgical, cement, mining, pharmaceutical, detergent and other particulate processing applications. However, these processes are challenging to model and operate because of the complex interactions affected by the particle parameters and process conditions such as particle size and distribution, shape, composition, fill level, rotation rate, and operating temperature.

Experiments are performed using 2 mm diameter silica beads to investigate the granular flow and heat transfer by conduction and radiation inside a 6-inch diameter and 3-inch-long stainless-steel rotary drum. The drum is attached to two 10-inch titanium wheels, hence preventing the direct contact of the drum wall with the rollers. One end of the drum is closed using a transparent quartz window, capable of handling high temperatures. Another end of the drum is closed using a sapphire window, compatible with an IR camera, specifically chosen to obtain a high transmittance to the infrared light. The drum walls are maintained at a desired high temperature using Ni-Cr resistive wires and a PID controller. The particle bed temperature evolution is recorded on the IR camera, which is used to determine the average particle bed temperature over time.

The current experimental research is focused on understanding the effect of low rotation rates (0 – 10 rpm) on heating by conduction and radiation at high temperatures (>600°C). Rotation rate plays a vital role in determining the contact time between the drum wall and the particles, which affects the amount of heat transferred by conduction and radiation in the rotary drum. This work will enable the determination of the contributions of conduction and radiation in heating the particles, and also serve to bridge the gap in the literature for experimental data and validating simulation-based research.

80d Comparative Study on the Collection Performance of a Coupled Cyclone with Built-in Circulating Granular Bed Filter for Geldart A and C Particles

Gao Sihong, Yiping Fan, Chunxi Lu

Fossil fuel, including coal, petroleum and natural gas, remains to be the most important energy supplying substance for the world. However, serious environmental problems have been brought by the combustion of massive fossil fuel, particularly, for the high temperature flue gas emission. The traditional separators including the cyclone separator generally fail to meet the emission requirements in the harsh operating conditions such as high temperature and high pressure (HTHP). For the development of new generation of gas purification techniques, diverse separation mechanisms are attempted to be coupled into one equipment. But until now, there are no reliable coupling separators in industrial applications meeting with the increasingly strict emission regulations and the comprehensive HTHP requirements. The HTHP gas purification technique has become a bottleneck of a number of advanced fossil fuel utilization processes, including the fluid catalytic cracking, the methanol to olefins process, the integrated gasification combined cycle and the biomass gasification etc.

A coupled cyclone separator with built-in circulating granular bed filter (C-CGBF) is idealized for integrating the benefits of centrifugation and filtration, in which arrangement the built-in granular bed is located around the core tube and in parallel with the cyclone shell, as shown in Fig. 1. In the cyclone field caused by the swirling flow, most of the large-diameter dust particles are firstly removed by the centrifugal force, which effectively reduces the filtration load of the following built-in granular bed; the small-size dust particles with a diameter less than 10 μm are then filtrated by the granular bed, so that the gas-solids separation efficiency can be further improved. Thus, the synergistic strengthening of the gas-solids separation can be achieved. In order to accomplish a continuous operation and realize the reuse of the collecting granules, a riser-spouted bed regenerator is adopted to regenerate and circulate the collecting granules.

The de-dust performance of the C-CGBF for two kinds of dust including the Geldart's Group A and C particles was comparatively investigated under a wide range of operating conditions including the inlet gas flowrate, the inlet dust concentration and the granules circulating flux in this work. The UOP 13X-APG adsorbents with an average diameter of 2.07 mm were packed in the granular bed as the collecting granules.

For both the Geldart's Group A and C dust particles, the experimental results shows that the collection efficiency can be improved by increasing the inlet dust concentration, while the pressure drop increases consequentially. High inlet gas flowrate leads to a decline of the collection efficiency and an increase of the pressure drop.

A real-time monitoring method is proposed for timely reflecting the instantaneous variation of the granule circulating flux, in which way a notable critical flux is observed. Under the moving bed operating regime, there exists a minimum critical granule circulating flux (W_{sc}), below which the operation regime tends to transform from the moving bed into the fixed bed. The W_{sc} for the Geldart's Group C particles is between 0.10~0.15 kg/s, while for the Geldart's Group A particles, the W_{sc} is between 0.04~0.10 kg/s. Increasing the granule circulating flux is benefit for the operation stability. However, the collection efficiency is decreased.

The contribution rations of cyclone shell to the collection efficiency of the C-CGBF system are all above 80% for both the Geldart's Group A and C dust particles. Increasing both the inlet dust concentration and the inlet gas flowrate can strengthen the centrifugal separation. Indicated by the size distribution of the captured dust particles, the cyclone shell can efficiently capture the particles with a diameter above 10 μm; while the particles captured by the built-in granular bed are generally smaller than 10 μm. Comparing the influences of different type of dust particle, the volume percentage (φ) of the particles with a diameter lower than 10 μm in the inlet dust has a significant effect on the downward moving of the collecting granules and the pressure drop of the C-CGBF. With an increase of φ, the growth rates of the pressure drop under the fixed bed and moving bed operating regimes are both increased. The collection efficiency is weakened and the critical granules circulating flux is also increased.

Based on the experimental data, comprehensive correlation equations for determining the collection efficiency and the pressure drop, with respect to the operating parameters and the properties of the dust particles, are obtained respectively. The errors between the calculated value and the measured data are all within ±10%. They are anticipated to be used in the engineering design and operation.

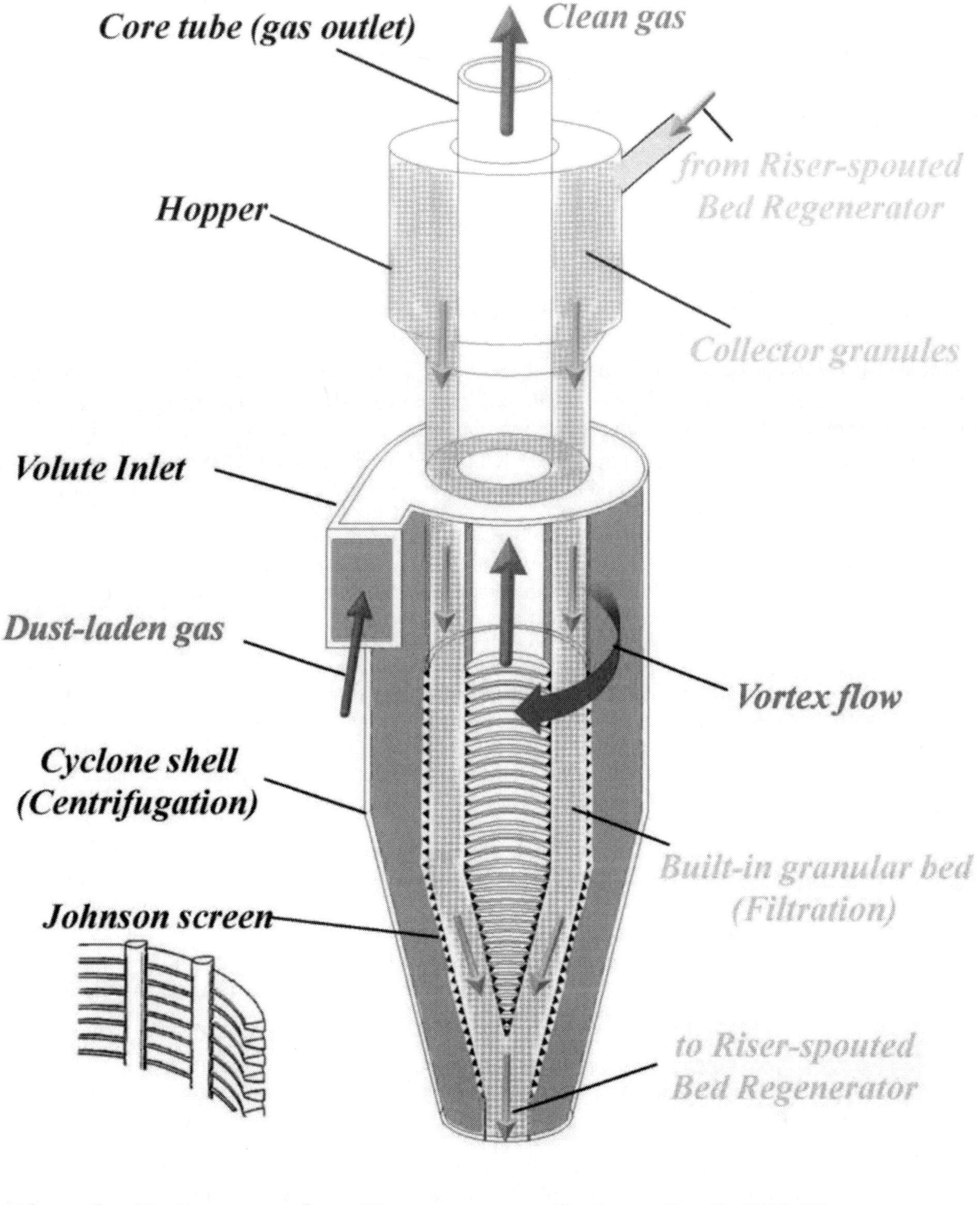

Fig. 1. Schematic diagram of the C-CGBF system

80e Predicting the Effect of Particle Shape on Random Packing: The Case of Nonequant Shapes

Anna Jaeggi, Anna-Maria Eckel, Ronny Pini, Ashwin Kumar Rajagopalan, Marco Mazzotti

The packing behaviour of granular materials is relevant to a variety of fields, for example filtration[1], packed beds[2], materials processing[3], to name a few. In all these applications, the accurate prediction of random packing structures based on a population of particles can facilitate more efficient processes, in terms of both the energy consumption and the productivity. For example, in filtration a low void fraction leads to a greater filter cake resistance, causing high pressure drop, and long filtration times.

The structure of a random packing of an ensemble of particles is affected by various factors, including surface properties, size distribution width, and particle shape. Previous work relating particle shape to packing densities has considered basic primary shapes[4], (e.g. spheres, cubes), as well as different convex and concave shapes[5] (e.g. stellated octahedra, tetrapods). A recent report has demonstrated the use of Monte Carlo simulations to predict the void fraction of packings of elongated particles of varying aspect ratio[6]. Despite the progress, several challenges related to the use of such models still exists, including the consideration of more complex shapes and more importantly their independent experimental validation.

The overarching goals of this work are (i) the systematic experimentally investigation of the effect of shape on packing in terms of aspect ratios, allowing for equant, elongated, plate-like particles, and everything in between; and (ii) the development of a model capable of predicting and explaining the porosity and packing structure for all the aforementioned shapes.

Here, we describe the development of a Monte Carlo framework to simulate the settling of particles of different sizes and shapes into a container. The settling process is simulated through stochastic, but preferentially downwards movement. Subsequently, we check if the particle's new position and orientation leads to intersection with other particles in the vicinity, the floor or the wall of the container. For computational efficiency, in the simulations we assumed all the particles to be spherosimplices. We considered several populations that have different aspect ratios (as shown in Figure 1).

We validate the trends observed in the Monte Carlo simulations through a systematic experimental campaign. To this aim, we use monodisperse 3D printed particles with the same shapes as the ones used in the simulation. In the experiments, the 3D printed particles were poured into containers and subsequently imaged with a medical computed tomography (CT) scanner, which was previously used for related purposes[7]. We developed an image analysis pipeline to extract quantifiable packing metrics from the tomographs. These metrics include the overall void fraction of the packing, vertical and radial porosity profiles, and particle orientations.

We performed several experiments to study: (a) the effect of the wall of the containers; (b) the repeatability of measurements; (c) the effect of tapping/vibrating the particles in the container; and (d) the effect of mixing particles of different shapes.

Based on the experimental and the computational outcome shown in Figure 1, we make two important observations. First, we observe that the porosity of the packing depends on the particle shape. Second, we show that the experimental data are in good qualitative agreement with the outcome from Monte Carlo simulations, despite the absence of any model parameter tuning. For all the particle shapes, the void fraction is slightly overestimated, the error being less than 0.05 in absolute terms. Additionally, we have also made other observations which are not shown here. First, as expected, vibration leads to a higher reduction in void fraction when applied to higher aspect ratio particles. Second, the magnitude of the wall effect depends on the particle shape. Third, the void fraction of populations with mixed particle shapes are averages of their corresponding constituent shapes.

To conclude, this work has led to a number of outcomes. First, we have a model that can qualitatively predict the packing behaviour of nonequant particles. Second, we have found a non-trivial dependence of packing behaviour (including the overall void fraction of poured and vibrated piles, and the wall effect) on the aspect ratio. In the future, this model can be used in a variety of real-world settings (e.g. pharmaceutical production chain) to reduce the experimental effort to determine packing behaviour.

References

[1] Murugesan, S., Sharma, P. K., andamp; Tabora, J. E. Design of Filtration and Drying Operations. *Chem. Eng. Pharm. Ind. Randamp;D to Manuf.* **2010**

[2] Achenbach, E. Heat and flow characteristics of packed beds. *Experimental Thermal and Fluid Science* **1995**, *10*(1), 17–27

[3] Dinger, D. R., andamp; Funk, J. E. Particle-packing phenomena and their application in materials processing. *MRS Bulletin* **1997**, *22*(12), 19–23.

[4] Li, B., Zhang, H., Saranteas, K., andamp; Henson, M. A. A rigid body dynamics model to predict the combined effects of particle size and shape on pressure filtration. *Sep. Purif. Technol.* **2020**, 119462.

[5] Landauer, J., Kuhn, M., Nasato, D. S., Foerst, P., andamp; Briesen, H. Particle shape matters – Using 3D printed particles to investigate fundamental particle and packing properties. *Powder Technol.* **2020**, *361*, 711–718.

[6] Perini, G., Avendaño, C., Hicks, W., Parsons, A. R., andamp; Vetter, T. Predicting filtration of needle-like crystals: A Monte Carlo simulation study of polydisperse packings of spherocylinders. *Chem. Eng. Sci.* **2021**, *230*, 116151.

[7] Pini, R., andamp; Madonna, C. Moving across scales: a quantitative assessment of X-ray CT to measure the porosity of rocks. *J. Porous Mater.* **2016**, *23*(2), 325–338.

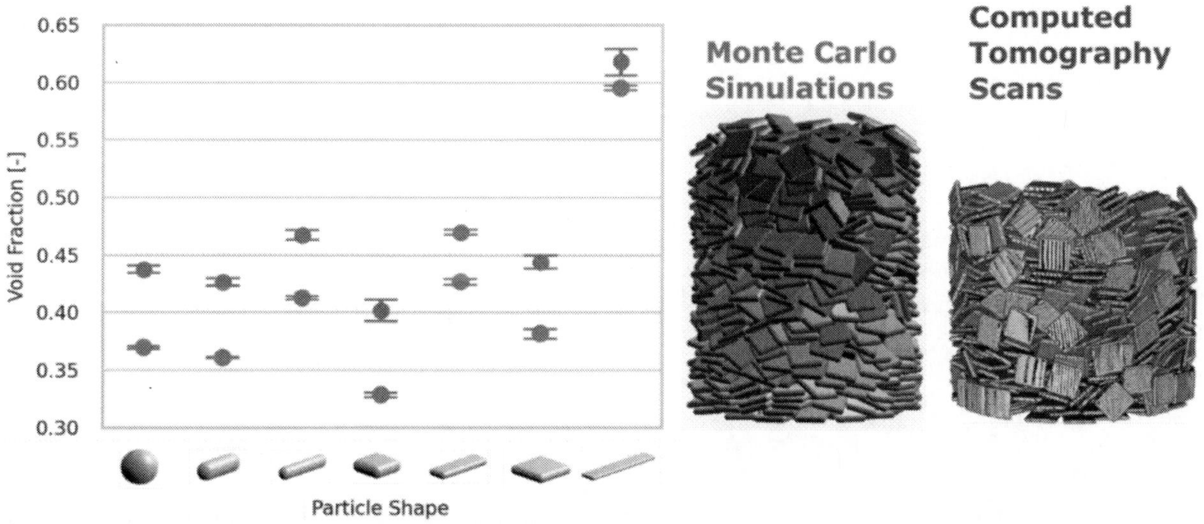

Figure 1: Void fraction of random packing as a function of particle shape obtained through experiments (gray) and simulations (red). The error bars indicate the standard deviation arising from repetitions of the experiments and simulations. Examples of piles obtained through Monte Carlo simulations (red) and through scanning the 3D printed particles using the medical CT scanner (gray) are shown on the right.

80f Powder Electrostatics: Focus on Fundamentals Needed to Understand Applications

Geoffroy Lumay, Eric Opsomer, Nicolas Preud'homme

When two particles are colliding or sliding on each other, electric charges could be exchanged at the surfaces. This contact electrification called triboelectric effect is an old fundamental scientific subject. However, despite many studies dedicated to this subject, the fundamental mechanisms behind the triboelectric effect are not fully understood. In particular, the charging of objects composed by the same isolating material and charging of particulate materials are still a matter of theoretical investigations. From an experimental point of view, difficulties are also emerging when an electrostatic measurement is performed on a particulate system.

In this presentation, we will look at the state of the art on powder electrostatics and we will analyze the recent theoretical models. Afterward we will make the connection with experimental and numerical (DEM simulations adapted to implement tribocharging) studies. From an experimental perspective, we will focus on a measurement method (GranuCharge from GranuTools) able to measure the ability of a powder to electrostatically charge during a flow.

80g Development of a Model for the Prediction of the Ignition Properties of Combustible Dusts Undergoing Homogeneous Combustion: Application to Sulfur Dust.

Jack Altwal, Tomasz Olewski, Luc Vechot

The determination of ignition properties such as the Minimum Explosible Concentration (MEC) and Minimum Ignition Energy (MIE) is critical for the control of the risks associated with combustible dust explosions. These properties are currently determined experimentally at laboratory scale using standardized tests. These tests are associated with several limitations: the tests require significant time and resources and the results are sample-specific. The particle size distribution of the samples may also get altered during testing by the equipment. This presentation describes a novel approach for the determination of the MEC and MIE of combustible dusts undergoing homogeneous combustion, using sulfur dust as a case study. The proposed model is based on the modeling of the heat and mass transfer phenomena occurring when a sulfur dust particle in a control volume is exposed to a planar hot surface acting as an ignition source. The model predicts the increase in temperature of the dust particle, its fusion, and vaporization in the control volume. An ignition criterion was developed on the basis of the accumulated amount of sulfur vapor and energy in the control volume. The model was used to predict the influence of the dust particle diameter on the MEC and MIE of combustible sulfur dust.

80h The 20-L Chamber and the Provision of Dust Explosion Data for Industrial Use

Paul Amyotte

The 20-L chamber has a 40-year history of widespread use for dust explosibility testing and research. Skjold (2018) describes its origin in the late 1970s when Richard Siwek demonstrated good agreement between his 20-L design and Wolfgang Bartknecht's 1-m^3 vessel in terms of comparable K_{St} (size-normalized maximum rate of pressure rise) values. There are now over a hundred 20-L chambers in operation globally.

The current presentation draws on the experience of the authors in providing test data to industry and conducting fundamental and applied research using the 20-L chamber. The objective is to elucidate key features that should be kept in mind both when operating this device and when analyzing the explosibility data it yields. Each of the items in the following list will be explained in detail with supporting examples and references to the technical and standards literature. The 20-L chamber:

- is a workhorse that has served the dust explosion community well for many years.
- has been used to produce an enormous amount of explosibility data.
- is more plentiful worldwide than the 1-m^3 chamber.
- has more than one design.
- must be operated following standardized protocols if test results are to be used for explosion prevention and protection.
- typically uses multi-point ignition sources.
- is capable of producing repeatable and reproducible results.
- can be used to test for a variety of explosibility parameters (e.g., maximum explosion pressure, size-normalized maximum rate of pressure rise, minimum explosible concentration, and limiting oxygen concentration).
- can be used to test a variety of materials (e.g., spherical or near-spherical dusts, nano-size dusts, flocculent dusts, and hybrid mixtures).
- turbulence levels decay with time (Mercer et al., 2001).
- dust concentrations are nominal (Kalejaiye et al., 2010).
- process of dust dispersion can alter the dust particle size distribution.
- can be overdriven or underdriven with respect to the 1-m^3 chamber.
- can be modelled.
- can yield results that might be surprising to the user (e.g., when testing nano-size dusts).
- is a significant improvement over the 1.2-L Hartmann bomb.
- is the best laboratory-scale alternative we have to the 1-m^3 chamber.

Overall conclusions on present and future use of the 20-L chamber will also be given in the presentation. Emphasis will be placed on the quality of the explosibility data produced by the 20-L chamber (especially for marginally explosible dusts), relative to the usefulness of the data for design of industrial prevention and mitigation measures.

References

Kalejaiye, O., Amyotte, P.R., Pegg, M.J. andamp; Cashdollar, K.L. (2010). Effectiveness of dust dispersion in the 20-L Siwek chamber. *Journal of Loss Prevention in the Process Industries*, 23(1):

46-59.

Mercer, D.B., Amyotte, P.R., Dupuis, D.J., Pegg, M.J., Dahoe, A., de Heij, W.B.C., Zevenbergen, J.F. andamp; Scarlett, B. (2001). The influence of injector design on the decay of pre-ignition turbulence in a spherical explosion chamber. *Journal of Loss Prevention in the Process Industries*, 14(4): 269-282.

Skjold, T. (2018). Dust explosion modeling: Status and prospects. *Particulate Science and Technology*, 36(4): 489-500.

335a Integration of High-Fidelity Simulation and Surrogate Modeling for Reduced Model Development in Continuous Pharmaceutical Unit Operations

Yingjie Chen, Pooja Bhalode, Marianthi Ierapetritou

Developing accurate prediction of continuous manufacturing (CM) lines has been an active research area for pharmaceutical industry with economic incentives and government initiatives to manufacture high quality products [1, 2]. Process models used to obtain the prediction for unit operations are commonly developed using system-level process knowledge and experimental data, which provide overall systemic information. However, this can limit the insight into the unit dynamics and hinder the prediction of important critical quality attributes [3, 4]. In CM routes for solid-based products, continuous powder blender plays a central role, providing mixing of ingredient powders. For blenders, the important quality attributes (CQA) include blend uniformity (BU) and residence time distribution (RTD) [5]. Powder holdup within the unit is another important attribute that impacts material flow in the blender, along with BU and RTD. To model this unit operation, a popularly used approach involves use of RTD models [6, 7] which lacks calculation of BU and holdup variability. Recent studies have explored the prediction of BU using compartment modeling strategies [8], but these models do not capture the variation of powder holdup along the blender [9] along its effects on CQAs [10]. To address these drawbacks and develop accurate models for unit operations, the proposed work aims at developing a reduced model based on integration of high-fidelity simulations, that can incorporate detailed unit dynamics, with surrogate models to reduce the computational complexity. This approach is demonstrated for a continuous powder blender.

The proposed work is divided into two parts. The first part focuses on developing a reduced model of the powder blender using high-fidelity discrete element simulations (DEM) [11]. Initially, the entire blender unit is simulated using DEM. Based on the observed flow of powder, the unit is broken down into the inlet section, multiple intermediate sections, and the outlet section. The dynamic information of each section, such as powder holdup, blend uniformity, and residence time, is extracted. This information is subsequently used to construct a reduced process model of the blender, incorporating powder flow and mixing along the different sections of the unit. The resulting process model thus incorporates axial mixing, radial mixing, and variable holdup profiles along the length of the blender. The second part of the work aims at improving the developed process model to incorporate the effects of varying process conditions on the degree of mixing and powder flow using a surrogate model. Physics-informed neural networks (PINN) is implemented as the surrogate model, given the flexibility of this model to incorporate physics within the construction of neural network [12]. PINN follows the general structure of an artificial neural network (ANN) [13], with a modified loss function to apply penalty to the predictions that violate physical constraints such as material balances and limits of holdup variations [12]. Therefore, the predictions that violate the constraints result in an increase in the loss function, allowing the training algorithm to favor the predictions that satisfy the physical constraints. The PINN model is integrated within the developed reduced model leading to accurate predictions. The developed process model for continuous blender can thus provide accurate predictions of with low computational cost, allowing the integration into process flowsheets and the Pharma 4.0 framework. This approach can also be generalized to other processes to better utilize physical knowledge, reduce computational complexity in process simulation, and improve predictability.

References:

1. Yu, L.X. and M. Kopcha, *The future of pharmaceutical quality and the path to get there.*

International Journal of Pharmaceutics, 2017. 528(1-2): p. 354-359.

2. O'Connor, T.F., Yu, L. X. andamp; Lee, S. L. , *Emerging technology: A key enabler for modernizing pharmaceutical manufacturing and advancing product quality.* . International Journal of Pharmaceutics, 2016. 509: p. 492-498.

3. Jourdan, N., et al., *Compartmental Modelling in chemical engineering: A critical review.* Chemical Engineering Science, 2019. 210.

4. Rogers, A. and M. Ierapetritou, *Challenges and opportunities in modeling pharmaceutical manufacturing processes.* Computers andamp; Chemical Engineering, 2015. 81: p. 32-39.

5. Engisch, W. and F. Muzzio, *Using Residence Time Distributions (RTDs) to Address the Traceability of Raw Materials in Continuous Pharmaceutical Manufacturing.* Journal of Pharmaceutical Innovation, 2016. 11(1): p. 64-81.

6. Escotet-Espinoza, M.S., et al., *Effect of tracer material properties on the residence time distribution (RTD) of continuous powder blending operations. Part I of II: Experimental evaluation.* Powder Technology, 2019. 342: p. 744-763.

7. Escotet-Espinoza, M.S., et al., *Effect of material properties on the residence time distribution (RTD) characterization of powder blending unit operations. Part II of II: Application of models.* Powder Technology, 2019. 344: p. 525-544.

8. Bhalode, P. and M. Ierapetritou, *Hybrid multi-zonal compartment modeling for continuous powder blending processes.* International Journal of Pharmaceutics, 2021. 602: p. 120643.

9. Vanarase, A.U., *Design, modeling and real-time monitoring of continuous powder mixing processes*. 2011.

10. Palmer, J., et al., *Mapping key process parameters to the performance of a continuous dry powder blender in a continuous direct compression system.* Powder Technology, 2020. 362: p. 659-670.

11. Ketterhagen, W.R., Ende, M. T. A. andamp; Hancock, B. C. , *Process modeling in the pharmaceutical industry using the discrete element method.* Journal of Pharmaceutical Sciences 2009. 98: p. 442–470.

12. Sampat, C., and Rohit Ramachandran, *Identification of Granule Growth Regimes in High Shear Wet Granulation Processes Using a Physics-Constrained Neural Network.* Processes 2021. 9(5): p. 737.

13. Raissi, M., P. Perdikaris, and G.E. Karniadakis, *Physics-informed neural networks: A deep learning framework for solving forward and inverse problems involving nonlinear partial differential equations.* Journal of Computational Physics, 2019. 378: p. 686-707.

335b Breakage of Elongated Crystals in an Agitated Filter Dryer

François S. Hallac, Frans L. Muller, Andrew Bayly

Pharmaceutical crystal size and morphology are important properties as they affect the performance of downstream processes, bulk properties such as density and flowability and ultimately product performance. The crystal size and shape distributions present at the end of a batch crystallization can change significantly during the subsequent filtration and drying procedures. The crystal morphology of active pharmaceutical ingredients is often 'needle like' with high aspect-ratio and understanding and predicting the breakage of these particles in agitated drying systems remains a challenging and active area of research. Computational methods, such as Distinct Element Modelling (DEM), have been used to simulate the flow of particle beds for various stress conditions [1] and a range of particle aspect ratios [2]. The contact forces between particles induces an internal stress which is responsible of particle breakage, with a predominance of the bending stress component [3]. Consequently, to predict the breakage behavior, the maximum internal bending stress experienced by individual crystals needs to be estimated.

In this study, a shear cell is built in DEM and mimics the stress environment experienced by particles in dryers using moving parallel walls and periodic boundaries. Elongated rigid particles are modeled with clumped spheres and experience stress due to the shear application in the box. The particles interactions are governed by Hertz-Midlin contact model with no cohesion to simulate a dry system. The internal stress of individual particles is calculated using Euler equation of motion and a particle internal stress distribution is obtained for different stress conditions (normal and shear) and particle physical properties. The Young's modulus and yield stress of a sample of Beta-Glutamic Acid (β-LGA) crystals are experimentally assessed with a novel 2-point cantilever bending method using Atomic Force Microscopy [4].

The particle internal stress distributions obtained from simulation are coupled with the experimental breakage strength distribution of the β-LGA sample to predict the breakage rate of the particles within the bed during the shearing phase. This breakage rate is shown to increase exponentially with both the normal stress acting on the shear layer and the particle elongation.

To predict the evolution of the shape and size distribution of dry particle in an agitated filter dryer estimates of normal stress in a dryer were used, along with the estimated breakage rates, in a simple population balance model. The results are compared to size and shape measurements from lab-scale agitated drying experiments on the studied sample of β-LGA crystals and show that the model captures the trends seen experimentally. The use of these approaches gives key insights on the causes of particle breakage and through their generalisation in the pharmaceutical industry they are expected to allow the optimisation of product quality during scale-up, while reducing the actual application of resource-consuming try-and-see methods.

[1] M. Boton et al., Physical Review E 87 (2013) 032206.

[2] Y. Guo et al., Physics of Fluids 25 (2013) 063304.

[3] B. Remy et al., AIChE Journal 61 (2015) 407-418.

[4] F.S. Hallac et al., CrystEngComm 21 (2019) 5738-5748.

335c DEM-Based Refill Strategy Optimization of a Twin-Screw Feeder

Peter Toson, Johannes G. Khinast

Residence time distribution (RTD) modeling has proven to be a valuable tool for material tracking in continuous pharmaceutical processes. Refilling is thoroughly studied in the literature, but the main focus lies on the feed rate disturbances. The impact of the feeder themselves on material tracking is often overlooked. Since the experimental methods to measure the RTD feeder discharging processes feeder are complex and material intensive, there is are only limited experimental RTD data available in the literature.

A DEM (discrete element method) simulation of a discharge of a Coperion KT-20 feeder in volumetric mode is available in an open access repository [1]. The results show that the simplest of material tracking assumptions – first in, first out (FIFO) – does not hold in this feeder because a large fraction of material that is moved and intermixed by the agitator. The DEM simulation shows that it is a much better approximation to assume discharge of a perfectly mixed material. In addition, the simulations show that there is a tendency to discharge material located above the agitator early. In order to predict the behavior during multiple refill events, three models in order of increasing complexity are presented:

(1) A simple reduced order model that assumes perfect intermixing of old and new material, leading to an exponential decay of the concentration of the old material in the hopper. This is a good approximation when discharging down to 30% before refilling (around the agitator axis)

(2) A RTD model based on the discharge of material layers. This model captures the preferred discharge of the top material layers, but fails at very early refills above 70% fill level.

(3) A DEM extrapolation method colloquially called 'Relay Race' [2]. This model accurately describes the trajectories of individual particles over multiple refills.

Refilling at fill levels below 30%, which corresponds to the agitator axis, can be accurately described by any of these models. However, cohesive powders often require refilling at higher fill levels in order to ensure a stable mass flow rate. In this case, discharge of material above the agitator axis plays an important role and one of the more complex refill models is then required.

References

[1] Toson and Khinast, 2019. *Particle-level residence time data in a twin-screw feeder*. Data in Brief, 27: 104672. doi:10.1016/j.dib.2019.104672. dataset doi: 10.17632/D76RZZD8R7.2

[2] Siegmann et al., 2021. *Massively speeding up DEM simulations of continuous processes using a DEM extrapolation*. Pow Tech, 390: 442–455. doi:10.1016/j.powtec.2021.05.067.

335d How Critical Is the Wall Cohesion for Flow of Cohesive Powders? the Cases of Tablet Press Feeder and Hopper Screw Feeder

Ankita Sharma, Lokeshwar Mahto, Tarun De, Jayanta Chakraborty, Anurag Tripathi, Jitendra Kumar, William Ketterhagen, Maitraye Sen

The cohesion between powder particles and process equipment walls are important factors affecting the flow of the bulk powder. Such effects are seen in experiments as well as in Discrete Element Method (DEM) simulations. However, most of the present DEM studies do not differentiate between the particle-particle and particle wall cohesion and identical values are assumed for both pairs. In this study we will show that it is important to assume appropriate and different values for these two kinds of cohesions and simulation results change significantly if this difference is ignored. We take the examples of the Tablet Press Feeder (TPF)[1] and the Hopper Screw Feeder (HSF)[2,3] to demonstrate these ideas.

DEM studies have been conducted for the TPF and HSF using LIGGGHTS public where SJKR2 (Simplified JKR) model has been used for cohesion. The particle-wall cohesion (P-W) was varied from zero to that of the particle-particle cohesion (P-P) and the effect on the process performance was studied. For the TPF, if the P-W cohesion is significantly smaller than the P-P cohesion, large variations in the mean tablet mass and relative standard deviation are observed, as shown in Figure 1(A). A similar effect has been observed for the HSF as well. In this case, even a highly cohesive system (with CED = 100,000 J/m3) can flow well if the wall cohesion is minimum. However, the flow will stop by bridge formation if the P-W and P-P cohesion values are of comparable magnitude, as shown in Figure 1(B). Many more interesting features are also revealed in these simulations which will be presented in this talk.

[1] W.R. Ketterhagen, 'Simulation of powder flow in a lab-scale tablet press feed frame: Effects of design and operating parameters on measures of tablet quality', Powder Technol. 275 (2015) 361–374. https://doi.org/10.1016/j.powtec.2015.01.073.

[2] J.W. Fernandez, P.W. Cleary, W. McBride, 'Effect of screw design on hopper drawdown of spherical particles in a horizontal screw feeder', Chem. Eng. Sci. 66 (2011) 5585–5601. https://doi.org/10.1016/j.ces.2011.07.043.

[3] Q.F. Hou, K.J. Dong, A.B. Yu, 'DEM study of the flow of cohesive particles in a screw feeder', Powder Technol. 256 (2014) 529–539. https://doi.org/10.1016/j.powtec.2014.01.062.

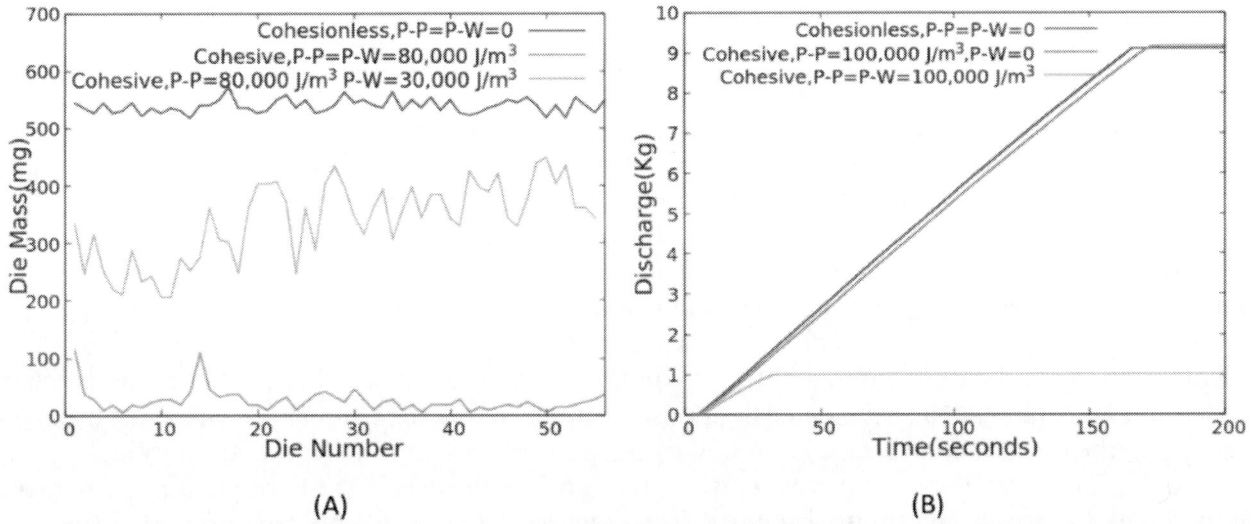

Figure 1: (A) Variation of die fill mass with die number in tablet press feeder at CED of 80,000 J/m^3. (B): Discharge through hopper screw feeder at CED = 100,000 J/m^3.

335e Application of Peridynamics to Predict Pharmaceutical Tablet Robustness

Sean Garner, William Ketterhagen, John C. Strong, Stewart Silling

In the advancement of drug product development, formulation design has evolved from more of a trial-and-error based approach to a science and physics-based approach that considers the chemical properties, the material science related properties, and mechanical properties of the powder blend. The ability to predict the manufacturability and performance of a given powder formulation greatly accelerates the development of formulation drug products. The anticipation and possible mitigation of tablet damage represents a specific area of interest in predicting formulation and tableting performance. As a result, practitioners of pharmaceutical science have turned to advanced modeling techniques. One such advanced modeling technique is the finite element-based continuum mechanics modeling approach that has been extensively used for predicting the mechanical behavior of powder material in the compaction processes. In this work, we seek to improve upon current modeling methodologies by developing a novel approach that utilizes the use of the Peridynamics, a numerical method with the ability to capture discontinuities, such as damage, in a new continuum mechanics approach. By utilizing the combination of the finite element method (FEM) and Peridynamics (PD) we aim to predict the evolution of damage and occurrences of failures in pharmaceutical tablets by linking (1) the finite element method – to elucidate the behavior of powders during compaction – with (2) the peridynamic modeling technique – to model the discontinuous nature of damage and predict compact breakage during the critical unloading/ejection stages. This work can be used to identify approaches to proactively modify material properties through formulation decisions and adjust processing conditions and/or tooling geometries to optimize compaction process efficiency and reduce compact breakage.

335f Tribocharging Quantification to Enhance Processability

Filip Francqui, Salvatore Pillitteri, Aurelien Neveu, Geoffroy Lumay

Powder flowability is an important parameter to control in most manufacturing processes. Powder conveying is a common step to transport the material from a process step to another. During this transport, the powder must exhibit good flow properties. However, particle to particle and particle to pipe rubbing can build up charge in the powder. Therefore, a powder that has a good flowability at the beginning can become cohesive due to the charge buildup and shows a lower flowability as the interactions between particles increase due to the Coulomb force. This charge buildup depends on the nature of the powder as well as the material of the pipes. Thus, it is important to be able to predict and to measure the chargeability of a powder for process improvement, by selecting the adequate powder or pipe material.

While the mechanisms controlling triboelectric charging are still poorly understood, it remains possible to evaluate the ability of a powder to gain charge by measuring its charge density after a flow. We present here results on the chargeability of hypromellose powders with different grades. The electrostatic charging has been investigated with the GranuCharge. It consists of a steel bend pipe, which reproduces most of possible charging ways, i.e. rubbing between grains or with the tube and collisions. The powder flows in the tube and falls in a faraday cup which measures the total electric charge of the powder. Knowing the mass of powder, one defines the charge density of the powder.

One can build a classification of the different powders according to their chargeability. This classification is then correlated to what can be observed for the flowability of the different powders with a loss-in-weight compact feeder. Moreover, we show the effect of the pipe material on the chargeability by using different tubes (Stainless Steel, PVC, Aluminium, ...) to highlight the importance of selecting the good material for conveying. These results highlight the impact of charge buildup during powder flow and the importance to characterize it for process improvement.

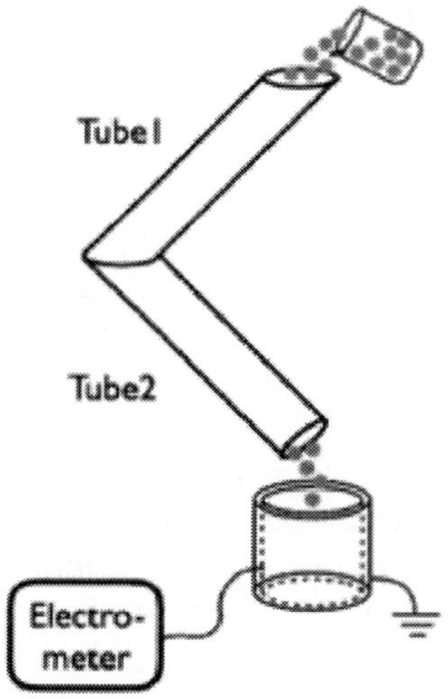

335g Pharmaceutical powder compaction; an accurate finite-element representation and prediction of binary mixtures

Danny van der Haven, Frederik H. Ørtoft, Kaisa Naelapää, Ioannis Fragkopoulos, James A. Elliott

Despite tablets being one of the most commonly used drug dosage forms, the design of a successful tablet production processes still involves a substantial element of trial and error. Tablets are typically made by the direct compression of pharmaceutical powders. However, if this process is not carefully controlled, the resulting tablets can be too weak, too strong, or show defects. This complex problem arises because the tablet's properties and occurrence of defects all depend on the 1) tablet geometry, 3) powder formulation, and 3) compression settings such as the maximum force.[1] New tools that can predict tablet quality depending on each of these production parameters are highly desirable. In this way, the time-to-market and development costs of new drugs could be reduced significantly.

The finite element method (FEM) is a widely used tool to predict tablet quality as a function of production parameters.[2] Although FEM simulation provides detailed information, it requires detailed parametrisation of the constitutive material model, which is typically done on a case-by-case basis. Moreover, the accuracy of the results is typically limited to a narrow range of pressures.

In this work, we present an automated and highly accurate FEM parametrisation framework for powder compaction.[3] The framework takes experimental data from a compaction simulator and tablet-crushing tests to parametrise the density-dependent Drucker-Prager cap (dDPC) constitutive model for the powder. By using several physical constraints and extrapolations, the framework is highly accurate and valid over a broad range of pressures. Validation test were performed on micro-crystalline cellulose (MCC), dibasic calcium phosphate dihydrate (DCPD), and three mixtures thereof. Simulated and experimental compaction curves showed a mean deviation of 2.5% of the maximum compaction pressure, only slightly more than the experimental variability of 1.2%, over a range from 0 to 500 MPa. Shear stress profiles from simulations also provided a plausible mechanism for the experimentally observed chipping defects.

The newly proposed framework already allows the use of FEM to anticipate the effect of tablet geometry and compaction settings on tablet quality. However, investigating the influence of the powder formulation remains challenging. The current framework requires experimental data of each and every powder mixture, making the entire process incredibly cumbersome for even a small number of components.

In this work, we also propose a set of mixing rules using the isostress assumption (i.e. that the stresses on all the individual components of the powder are the same). The mixing rules were then be used to determine the FEM parametrisation of several mixtures using only the FEM parameters of the individual pure components. Three powder mixtures of MCC and DCPD were used for validation. The predicted and experimental compaction curves showed a mean deviation of 4.8% of the maximum compaction pressure. The predicted shear stress profiles were similar to those previously simulated, providing the same mechanism for the chipping defects. The mixing rules can thus be used to estimate the influence of the powder formulation on the tablet quality.

Finally, the effect of mixture homogeneity was investigated. Because the mixing rules using the isostress assumption gave accurate predictions, a powder distribution that gives a similar stress distribution for all materials should also result in accurate predictions. Simulating tablets in which MCC and DCPD were separated into two horizontally stacked layers resulted in a mean 4.5% error with respect to the experimental homogeneous mixtures. This again justifies the use of the isostress assumption. A further comparison of spatial arrangements of the powder (vertical layers and a staircase pattern) showed that the residual radial pressure, a key indicator of tablet defects, was lower when the softer material (MCC) was placed near the die wall.

To conclude, the framework can be used to obtain accurate FEM parameterisations for powder compaction simulations. The framework also was extended with a number of mixing rules. Using either experimental data or mixing rules resulted in accurate predictions of the compaction curve and a similar chipping defect mechanism. In the future, this framework can be used to systematically asses the influences of tablet geometry, compaction force, and powder formulation on relative tablet quality, using only experimental data of the constituent single-component powders.

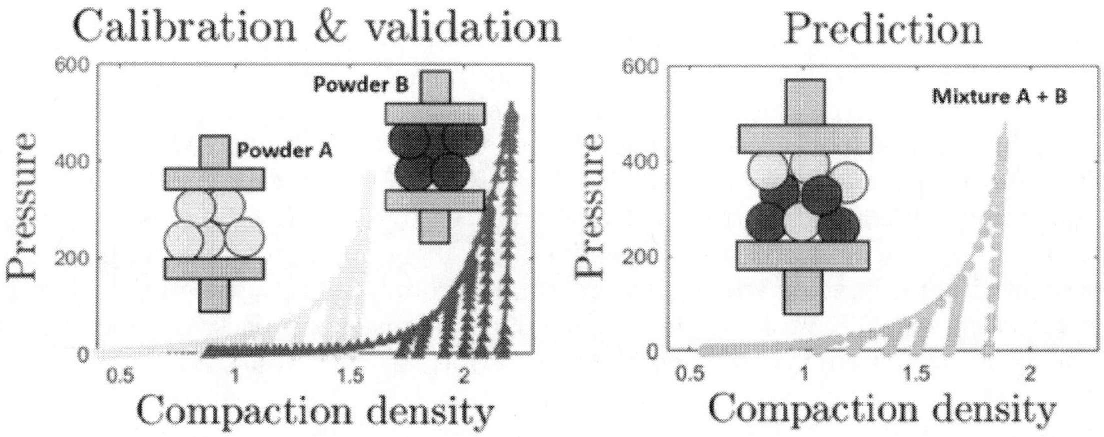

Figure 1: Graphical abstract of the presented work. Figure adapted from van der Haven et al.[3]

References

[1] Rana, Abhinav Singh, and SL Hari Kumar. "Manufacturing defects of tablets-a review." *Journal of Drug Delivery and Therapeutics* 3.6 (2013): 200-206.

[2] Partheniadis, Ioannis, Vasiliki Terzi, and Ioannis Nikolakakis. "Finite Element Analysis and Modeling in Pharmaceutical Tableting." *Pharmaceutics* 14.3 (2022): 673.

[3] van der Haven, Dingeman LH, et al. "Predictive modelling of powder compaction for binary mixtures using the finite element method." *Powder Technology* 403 (2022): 117381.

335h An Assessment of Batch Powder Bin Blending Performance Via Process Analytical Technology (PAT) and Discrete Element Method (DEM) Simulation to Predict Blend Homogeneity

William Ketterhagen, Greg Doddridge, Jeffery Larson, Leo Manley

In recent years, continuous mixing of powders has been a topic of great interest in the pharmaceutical industry. Despite these advances, many solid oral dosage manufacturing processes still rely on batch powder blending to mix incoming raw ingredients. In this work, we describe a comprehensive assessment of bin blending performance/blend homogeneity via calibrated DEM simulations and experiments with an acetaminophen formulation. A range of process conditions are examined including bin scale, rotation speed, and powder fill level. Various blend performance assessment techniques are utilized including on-line NIR and off-line NIR of powder thief samples. Each of these are compared with corresponding virtual powder samples from DEM simulations as well as a DEM system-wide mixing index. Several key points related to the impact of process parameters and sampling/analysis methods on the observed blending performance are described.

536a A Revolutionary, Evolutionary Approach to Particle Characterisation

Kit Windows-Yule

There exist a wide array of commercial powder characterisation tools capable of measuring a diverse assortment of bulk powder properties, such as flowability or cohesive index. Many of these properties are, however, purely heuristic and offer little physical insight into the micromechanical properties and behaviours of the materials characterised. This issue is troublesome not only in that it inhibits our understanding of particulate systems of interest, but also means that numerical models thereof (e.g. discrete element method simulations) which require details knowledge of powders' *microscopic* (particle-level) properties cannot be calibrated using commercially-available devices. While in academia researchers have created custom systems for the measurement of particle properties, these ad hoc approaches are not viable for industry.

Without proper calibration, DEM simulations are prone to being inaccurate, or even unphysical, thus preventing their widespread adoption in industry, where they otherwise stand to prove transformative to the design and optimisation of industrial particle-handling systems. In short, the lack of an efficient, effective manner for determining particles' micromechanical properties represents perhaps the most significant bottleneck in the industrial use of DEM, preventing the advent of a potentially paradigm-shifting move toward Industry 4.0 within the particle-handling industries.

The same evolutionary strategy may also be used to determine novel equations ... micro macro mapping

536c Effects of Particle Size, Shape and Moisture Content on the Flowability of Microcrystalline Cellulose Powder

Jordan Monroe, Heather Emady

Powders are encountered across multiple industries such as agricultural, cosmetics, food, mining, and pharmaceuticals. An important factor in the manufacturing process of pharmaceuticals is the flowability of powders to ensure uniformity in the feed out of hoppers and into tablets so that a consistent drug content can be maintained. Therefore, it is necessary to control and optimize factors affecting the flow of materials for efficient processing. The primary parameters that affect the flow of a powder include particle size, particle size distribution (PSD), particle shape, and moisture content. This work focuses on investigating the effects of particle size, moisture content, and particle shape on the flowability of microcrystalline cellulose (MCC), a common pharmaceutical excipient, using the FT4 Powder Rheometer. The study included particle sizes between 10 – 150 microns and moisture levels from 0 – 50% by weight. Using particle size and shape measurements from the Malvern Morphologi G3 particle analyzer, particle size, shape, and moisture content were correlated with the FT4 flowability results. This is the first study, to our knowledge, to quantify the effects of particle shape, size, and moisture content on the flowability of powders.

536d A Comparative Evaluation of Ca-Based Wastes for a Oxy-Fuel Circulating Fluidized Bed Combustor

See Hoon Lee, Hyun Jun Park, Chul Seung Jung, Jung Min Sohn, Sang Mun Jeong

Limestone ($CaCO_3$) particles are used for SO_2 absorbants or CO_2 carriers for various circulating fluidized bed (CFB) processes, such as oxy-fuel combustion, CO_2 capture process, and chemical looping hydrogen reactors. In oxy-fuel combustions, limestone directly injected into CFB risers combines with SO_2 to produce calcium sulfate ($CaSO_4$) via direct desulfurization reaction or indirect desulfurization reaction. In this study, Ca-based wastes such as eggshells or seashells was investigated as the alternatives of limestone particles. Since the loop-seal in oxy-fuel CFB processes is generally operated under CO_2 rich conditions, calcined Ca-based wastes are exposed to high CO_2 partial pressure conditions through the recirculation section for an extended period and re-injected into the riser as recarbonated wastes. To evaluate the usage of Ca-based wastes for oxy-fuel CFB combustions, their reactivities of calcination, recarbonization, and direct desulfurization were studied in a thermogravity analyzer (TGA) under oxy-fuel combustion conditions. Also, the morphological changes of Ca-based wastes were evaluated using XRD, FE-SEM, and other analysis methods. The TGA experiments show that recarbonation contributes to $CaSO_4$ conversion. Increasing the desulfurization efficiency using recarbonation can reduce the fixed investment and operating costs of oxy-fuel CFB plants because only desulfurization in the furnace is able to meet SO_2 emission regulations or lower the flue gas desulfurization (FGD) dependence.

536f Scale-up of Drying of Supported Catalysts in a Fluidized Bed

Carlin Leung, Justin Adler, Tim A. G. Langrish, Benjamin Glasser

Drying of supported catalysts is an essential unit operation in the manufacturing of catalysts. Several different types of dryers are used for catalyst drying include belt dryers, rotary kiln dryers, and fluidized bed dryers. In fluidized bed drying, the heated gas flows through a bed of solids and fluidizes the bed, promoting high rates of heat and mass transfer between the gas and solids. Fluidized bed dryers allow for simultaneous spray impregnation and drying of supported catalysts, while promoting high moisture content uniformity through fluidization. Although several studies have been done on fluidized bed drying, general scale-up equations for fluidized bed drying of catalysts are lacking, as drying in a fluidized bed depends on the material properties and fluidization conditions. Thus, it is difficult to choose the operating conditions to optimize the drying process due to several phenomena occurring simultaneously, including heat and mass transfer, drying and fluidization. In general, one would like to have sufficient understanding of the process so that experiments done on a laboratory scale could be used to predict behavior at larger scales.

Experiments were carried out in a Glatt GPCG-1 fluidized bed dryer. Wetted catalyst supports were used as the experimental material. The rate of drying can be described by the change in moisture content of the bed over time. Experiments were done for different temperatures, gas flowrates and bed loadings. Model equations for predicting the moisture content of the bed were implemented using reference drying curves. We found that the model was generally capable of predicting the moisture content of the bed and the rate of drying for various process conditions. Moreover, for different fluidization conditions, we found that the model was capable of predicting the time to dry to 0% moisture content for cases with a delay in fluidization. We found that the model worked best when the majority of the drying took place in the unhindered drying period where the rate was roughly constant. However even when drying took place in the hindered drying period, the model was still able to make reasonably good predictions. We also found that there was better agreement between the model and experiments for high bed loadings where the number of transfer units was large. This work provides quantitative findings on how process conditions affect the rate of drying for a bed of catalyst supports with corresponding scale up equations.

536h Conduction and Radiation Heat Transfer in a Rotary Drum

Bhaumik Bheda, Heather Emady

Rotary drums deliver high heat and mass transfer rates, making them extensively useful for powder mixing and heating processes in metallurgical, cement, mining, pharmaceutical, detergent and other particulate processing applications. However, these processes are challenging to model and operate because of the complex interactions affected by the particle parameters and process conditions such as particle size and distribution, shape, composition, fill level, rotation rate, and operating temperature.

Experiments are performed using 1-4 mm diameter silica beads at fill levels from 10–25% and rotation rates from 0–10 rpm. The granular flow and heat transfer by conduction and radiation are investigated inside a 6-inch diameter and 3-inch-long stainless-steel rotary drum. The drum is attached to two 10-inch titanium wheels, hence preventing the direct contact of the drum wall with the rollers. One end of the drum is closed using a transparent quartz window, capable of handling high temperatures. Another end of the drum is closed using a sapphire window, compatible with an IR camera, specifically chosen to obtain a high transmittance to the infrared light. The drum walls are maintained at a desired high temperature using Ni-Cr resistive wires and a PID controller. The particle bed temperature evolution is recorded on the IR camera, which is used to determine the average particle bed temperature over time.

The current experimental research is focused on understanding the effect of particle size, fill level, rotation rates, and their interactions on heating by conduction and radiation at high temperatures (>600°C). This work will enable the determination of the optimal operating conditions for a rotary drum for improving energy efficiency of the process. It will also serve to bridge the gap in the literature for experimental data and validating simulation-based research.

536p Optimization of Low Dose Powder Filling Process for Dry Powder Inhalers

Tanu Mehta, Zhanjie Liu, Bruhal Shah, Justin P. Lacombe, Bodhisattwa Chaudhuri

Purpose:

Dry powder inhaler (DPI) manufacturing involves handling low dose formulations. There are multiple methods exist for low dose DPI filling such as tamp filling, dosator filling, vacuum dosator filling and vacuum drum filling but powder filling in milligram fill weight is still a huge challenge.

As most of the pharmaceutical powders are poor flowing and show unpredictable flow behavior, this filling process becomes even more difficult and challenging. In this study, we have evaluated lab scale and pilot scale vacuum drum filling system by Harro Hofliger to understand the critical material attributes (powder density, particle size distribution and flowability) and critical process parameters (vacuum fill pressure, drum bore size, and stirrer type) on fill weight and %RSD for the DPI capsule filling process. The overall goal of this study was to optimize the vacuum drum filling process.

Methods:

During this study, pharmaceutical powders with different material properties were selected (table 1) and filled in size 3 capsules using Harro Hofliger's DrumLab (lab scale) and ModuC LS (pilot scale). Among the selected powders Lactohale 300/LH300, Inhalac 230 and Inhalac 120 are poor, fair and good flowing powders respectively. We have also evaluated an in-house formulation A. unlike other selected materials, proprietary formulation A was highly cohesive and low-density (<100g/l) demonstrating very poor flow properties. The selected powders materials were filled at different vacuum pressure levels using drum of bore size 7.5mm^3 and 30mm^3, and net fill weight of each capsule was recorded to study fill weight variation (%RSD) and product density (mg/mm^3). Additionally, the impact of stirrer geometry on fill weight and %RSD was also studied using two types of stirrers including standard stirrer and asymmetric stirrer. The standard stirrer is commonly used to fill good to fair flowing powders whereas asymmetric stirrer is suitable for poor flowing and cohesive powders.

Results:

The fill weight was a function of vacuum pressure and linear relationship was observed between fill weight and vacuum pressure (table 2) for all materials except Inhalac 120 as it is a good flowing powder. However, higher %RSD was observed with LH300 and formulation A at lower fill levels especially with standard stirrer. The asymmetric stirrer led to lower %RSD and higher fill weight compared to standard stirrer for LH300 (figure 1) and formulation A which could be attributed to its paddle shaped geometry.

Conclusions:

Vacuum drum filling is an efficient method for low dose powder filling and can be easily scaled from laboratory to pilot scale. However, it is important to understand the material attributes to determine process parameters such as vacuum pressure, drum bore size and type of stirrer to optimize overall filling process.

Keywords:

Low-dose filling, Drum filler, DrumLab

Material	Tap Density(g/l)	Bulk density(g/l)	D50(µm)	Compressibility Index(%)
Lactohale/LH300	520	260	<5	>25
Inhalac 230	850	700	70-110	18
Inhalac 120	830	720	110-155	13
Formulation A	-	90.44	-	43.4

Table 1: Material attributes of selected powders

Material	Vacuum(mbar)	Standard Stirrer		Asymmetric Stirrer	
		Fill weight (mg)	%RSD	Fill weight (mg)	%RSD
LH300	200	11.031	7.353	11.261	7.370
	400	13.531	7.499	14.453	3.774
	600	14.846	5.321	15.222	3.442

Table 2: Fill weight and %RSD for LH300 using $30mm^3$ bore size drum

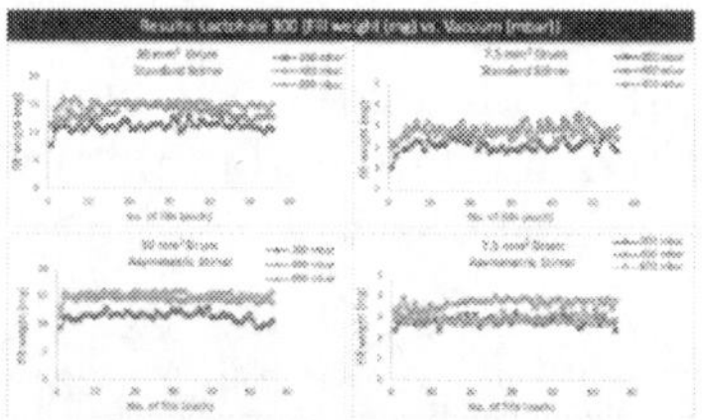

Figure 1: Filling data obtained for Lactohale 300 using DrumLab

536j Magnetic Hyperthermia in Biosimilar Colonic Tumor Environment Using Silica-Coated Superparamagnetic Iron Oxide Nanoparticles

Yuming Zhang, Alexandra Teleki

Colorectal cancer (CRC) is the fourth most deadly cancer in the world. Standard treatment options include surgery, chemotherapy and radiotherapy. In the clinic, those treatments are often combined or used with other alternative therapies to improve the treatment outcome. Magnetic nanoparticle hyperthermia has attracted clinical interest as an alternative treatment option for chemo-resistant cancer due to its advantages in deep tumor penetration, local tumor heating, and targeted tumor killing when functionalized with specific targeting moieties. This therapy relies on delivering superparamagnetic iron oxide nanoparticles (SPIONs) to the tumor site where they can generate heat in the presence of an alternating magnetic field (AMF). The local temperature increase causes cancer cell death by apoptosis. Current challenges of magnetic hyperthermia are to improve hyperthermia efficacy while minimizing any toxic off-target effects of the SPIONs. The former can be achieved by increasing the heat dissipation rate of SPIONs which is controlled by SPION size, composition and shape. Furthermore, the SPION heating efficiency also strongly depends on the composition and viscosity of the surrounding biological environment and will vary due to the dynamic biotransformations of SPIONs in vivo. In this study, we aim to optimize the hyperthermia effiency of orally administered SPIONs for local treatment of CRC. We study the hyperthermia efficiency of SPIONs in vitro in biosimilar colonic environments in order to predict the optimal subcellular environment for hyperthermia therapy of CRC. SPIONs were synthesized using a flame aerosol reactor that allows scalable manufacture of functional nanoparticles. We varied the SPION composition by addition of dopants (zinc, cobalt, manganese, gadolinium) that are known to enhance the saturation magnetization and hypethermia efficiency of SPIONs. The particles were characterized by nitrogen adsorption, X-ray diffraction, and their hydrodynamic diameter by dynamic light scattering. The crystallite size of particles was finely tuned to 15 nm by controlling flame process parameters to render them superparamagnetic. The particles were hermetically encapsulated by nanothin SiO_2 by in-situ flame coating to facilitate subsequent surface functionalization [1, 2]. Methoxy polyethylene glycol (mPEG) was successfully conjugated to the nanoparticle surface post-synthesis to improve nanoparticle suspension stability in biologically relevant media. The successful pegylation was confirmed by C-H stretching (2871 cm^{-1}) and bending (1454 and 1349 cm^{-1}) bands appearing in Fourier transform infrared spectra. The hydrodynamic diameter of mPEG-SPIONs in cell culture media decreased compared to the bare particles (e.g. the hydrodynamic diameter of SiO_2-coated $Mn_{0.6}Zn_{0.4}Fe_2O_4$ in cell culture media was 800 nm and 200 nm before and after pegylation, respectively). The mPEG-SPION suspensions were stable in cell culture media with minimal sedimentation even after 1 hour of AMF exposure. The heating efficiency of nanoparticles was characterized in biosimilar colonic tumor environments (biosimilar mucus, agar tumor phantom, and artificial colonic fluid) to study the outcome of CRC hyperthermia therapy *in vitro*. Preliminary data suggests that the hyperthermia efficiency is lower in highly viscous environments (e.g. biosimilar mucus and agar tumor phantom) compared to water. Overall, this study will allow the systematic development of surface-modified SPIONs for targeted and efficient hyperthermia therapy of CRC.

1. Teleki, A., et al., *Hermetically Coated Superparamagnetic Fe2O3 Particles with SiO2 Nanofilms.* Chemistry of Materials, 2009. **21**(10): p. 2094-2100.

2. Teleki, A., et al., *Highly scalable production of uniformly-coated superparamagnetic nanoparticles for triggered drug release from alginate hydrogels.* RSC Advances, 2016. **6**(26): p. 21503-21510.

536k Bioactivity of Flame-Made Calcium Phosphate Bionanomaterials for Bone Tissue Engineering

Yael del Carmen Suárez López, Georgios Sotiriou

The most common cause for bone implant failure is the poor biological function due to insufficient bone formation at the biomaterial-tissue interface and consequently, lack of osseointegration (Zhong et al 2018). Therefore, the overall trends for the development of bone implant synthetic biomaterials are shifting towards bioactive materials which resemble natural bone tissue to promote osseointegration. Natural bone tissue is porous, three-dimensional, and consists of type I collagen fibrils and calcium phosphate nanocrystals. To promote osseointegration of artificial materials, calcium phosphate nanoparticles can be synthetized in the laboratory through several wet strategies, but these techniques are often limited by their low yields and the resulting nanoparticles' lack of similar morphology to natural calcium phosphate (Nasiri et al 2018).

Therefore, single-step flame spray pyrolysis (FSP) has emerged as a technique for the synthesis of calcium phosphate fractal-like nanoaggregates that consist of spherical calcium phosphate nanoparticles with tuneable physicochemical properties, and with morphology that resembles the natural structure of calcium phosphate in bone tissue.

In this work, calcium phosphate nanoparticles that have different crystallinities, from hydroxyapatite to amorphous calcium phosphate, as well as different sizes and carbonate content, are synthesized by FSP (Figure 1a, 1b). The synthesized nanoparticles can be directly deposited on substrates in-situ during production, or the nanoparticles can be also integrated into polymer electrospun fibers, which mimic the natural bone tissue structure. The bioactivity of the synthesized materials by their cell-free biomineralization capacity was studied as well (Figure 1c, 1d). Finally, the materials could be used to promote stem cell differentiation towards osteoblasts without the need of medium supplementation (Schneider et al 2008). All the materials tested were able to induce cell-free biomineralization, although the rate of deposition and degree of crystallinity were dependent on the identity of the starting material. The results obtained here demonstrate the facile synthesis of calcium phosphate nanoparticles by FSP, the importance of tuning their physicochemical characteristics, and the potential to use these bionanomaterials for the promotion of osseointegration of synthetic implants.

References

Nasiri N, Mukherjee S, Panneerselvan A, Nisbet DR, Tricoli A (2018) ACS Applied Materials and Interfaces

Schneider OD, Loher S, Brunner TJ, Uebersax L, Simonet M, Grass RN, Merkle HP, Stark WJ (2008). Journal of Biomedical Materials Research - Part B Applied Biomaterials.

Zhong L, Qu Y, Shi K, Chu B, Lei M, Huang K, Gu Y, Qian Z (2018). Science China Chemistry.

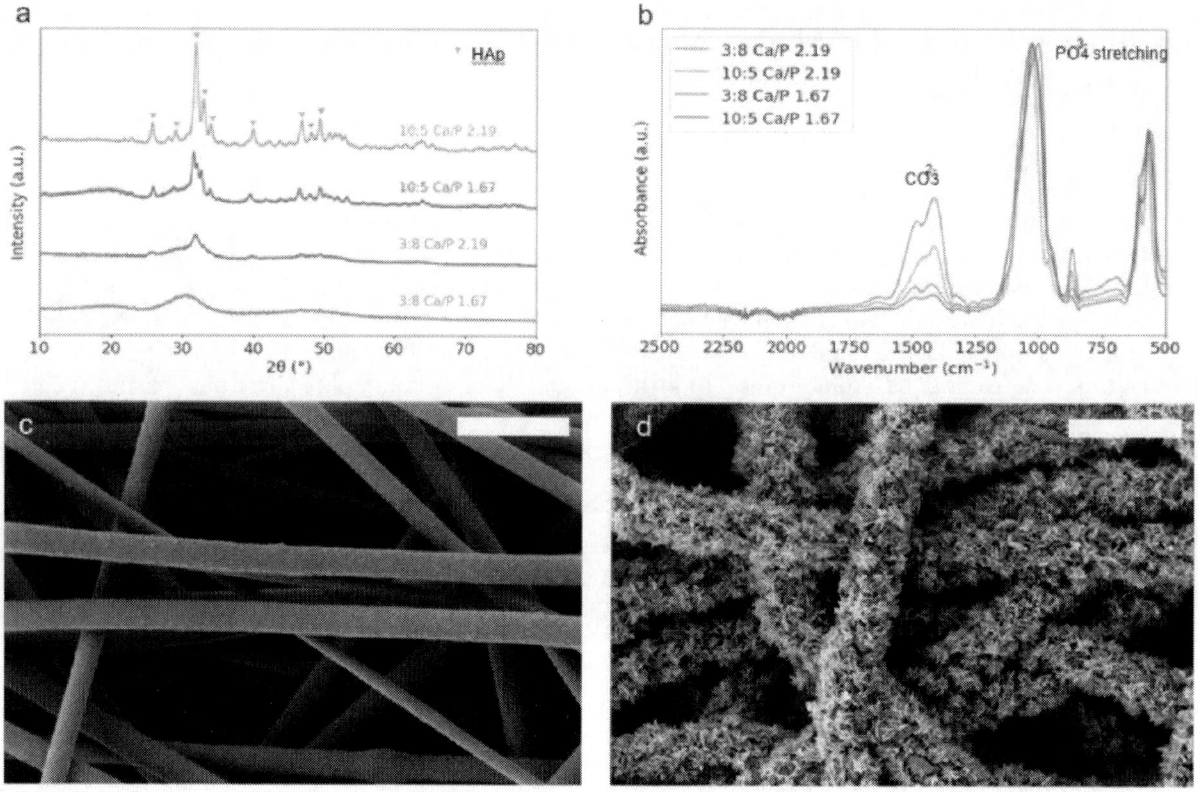

Figure 1. a) XRD spectra of four different nanoparticles produced by FSP with different flame conditions, where X:Y represent the precursor feed rate in mL/min (X) and oxygen dispersion gas flow rate in L/h (Y), and Ca/P represents the calcium to phosphate ratio of the nanoparticles produced. b) FT-IR spectra of the nanoparticles. c) Electrospun polymer fibers embedded with nanoparticles. d) Electrospun polymer fibers after 100 h of immersion in simulated body fluid. Scale bar is 2μm.

536l Manufacture of Complex-Shaped Tungsten Materials Via Atomic Layer Deposition and Direct Ink Writing

Hailey Loehde-Woolard, Elena Napoletano, Davis R. Conklin, Bergen Evans, James Smay, Alan Weimer

Tungsten (W) is a refractory metal that exhibits a unique combination of high melting point, strength, ductility, thermal conductivity, and corrosion resistance. As such, it is integral for use in extremely high temperature applications such as nuclear thermal propulsion and shielding elements of solar probes. However, due to the high melting point (3422°C), W is extremely difficult to melt and cast into various complex shapes. Most W based alloys are made through powder metallurgy, pressing and sintering W powder with the addition of transition metal sintering aids. This limits the complexity of the final fabricated part, limiting the ability to produce intricate components for the desired applications. Current state of the art additive manufacturing techniques for W and W alloys are powder bed fusion and binder jet printing. Both techniques are highly dependent on process conditions, which can introduce flaws such as stress gradients, porosity, and non-uniform shrinkage in the final part.

In this work we propose the use of direct ink writing (DIW) of W and W alloys to fabricate complex shaped components with near full density, high yield strength, and low surface roughness. Further, we utilize atomic layer deposition (ALD) to deposit sintering aids and dopants at low loading directly on the particle surface. A tungsten-based colloidal ink was formulated with an optimized solids fraction volume loading. The ink was then continuously extruded through a nozzle which enables bottom-up fabrication of the desired geometry, such as cylinders or lattices for testing. A design of experiments (DOE) was employed to optimize printing parameters in order to mitigate surface roughness and maintain print uniformity. The green bodies were dried and sintered at 1600°C in a low hydrogen atmosphere. Surface profilometry was used to obtain the sintered surface roughness and evaluate the DOE. The optimized printing parameters were employed to manufacture samples with various sintering aids and dopants, added through conventional mixing or ALD on the powder prior to ink formation. A secondary DOE was employed to evaluate the parameters that affect the final print density. The systematic optimization of the colloidal ink resulted in parts that demonstrated both densification and stability in complex geometry. Addition of the sintering aids promoted densification further at the lower sintering temperature and ALD was shown to be a successful loading method. This work introduces the use of ALD-functionalized W in colloidal gels for the purpose of DIW complex shapes.

536m Exploring $LaFeO_3$ oxygen Carriers for Reactivity Enhancement through Structural Changes in Chemical Looping Partial Oxidation System

Anuj Joshi, Sonu Kumar, Lang Qin, Zhuo Cheng, Hendrik Colijn, Zain Mohammad, Liang-Shih Fan

Improvements in the utilization of hydrocarbon feed to produce syngas, a valuable intermediate, have always been sought-after in the industry. Chemical looping partial oxidation (CL-POX) is a promising pathway for syngas production due to its superior reaction thermodynamics and energy efficiencies compared to conventional technologies. However, the success of CL-POX lies in its oxygen carrier, as it provides the oxygen required for the reaction, and controls the product yield. Several micro-sized oxygen carriers have been developed for conducting the CL-POX reaction. However, because of their low specific area and pore volume, the reaction kinetics tend to get suppressed. Thus, there is a research thrust in developing nano-sized oxygen carriers to enhance the kinetics.

In this work, $LaFeO_3$ nanoparticles were synthesized for carrying out the CL-POX reaction by embedding them in mesoporous SBA-15 support ($LaFeO_3$@SBA-15). For comparison micro-sized $LaFeO_3$ sample impregnated on SiO_2 support ($LaFeO_3$@SiO_2) was also synthesized. Thermogravimetric experiments (TGA) employing methane as reducing gas showcased that $LaFeO_3$@SBA-15 maintained stable performance over 15 redox cycles with nearly 1000% improvement in reaction rate compared to $LaFeO_3$@SiO_2. In addition, characterization techniques such as X-ray diffraction (XRD), nitrogen physisorption, and high-resolution transmission electron microscopy (HR-TEM) were used to investigate this dramatic increase in reaction rate. The analysis revealed that $LaFeO_3$ forms a different crystal structure when embedded in SBA-15 as opposed to SiO_2. This can be attributed to the mesoporous, tubular, and ordered structure of SBA-15 which enabled the formation of nano-scaled particles and stabilized a more active $LaFeO_3$ crystal structure. To corroborate the experimental results and explore the reaction mechanism over the different crystal structures, atomistic level-based density functional theory calculations were carried out. The findings from this study provide vital insights into the role of crystal structure in reaction kinetics and thus can be leveraged to develop highly active chemical looping oxygen carriers.

536n Refractory Tungsten ALD Coatings for Nuclear Thermal Propulsion Fuel Elements

Davis R. Conklin, Sarah Bull, Christopher G. McKinney, Jamelle K.P. Williams, Arne Croell, Jhonathan Rosales, Alan Weimer

Crewed missions to Mars will require engines with high specific impulse to achieve sufficient fuel efficiency for a return trip. Nuclear thermal propulsion (NTP) engines can achieve double the specific impulse of conventional chemical rockets by utilizing fission reactions to heat hydrogen propellant to around 2700 K before it is expelled through a supersonic nozzle. Fuel elements for NTP engines are typically composed of uranium oxide fuel pellets embedded in a refractory metal matrix and are susceptible to degradation at the temperatures required for NTP. Applying a barrier coating to the fuel pellets before fabrication of the cermet (ceramic-metallic) fuel element has been demonstrated to limit uranium hydride formation, prevent segregation of uranium oxide within the refractory metal matrix, and improve the final density of the fabricated cermet, all of which improve the durability of the fuel element. However, there is a need for finer control over the coating-fuel interface and an improved understanding of how coating properties influence performance in high-temperature, hydrogen-rich environments.

In this work, we utilize particle atomic layer deposition (ALD) to synthesize a tungsten coating on yttria-stabilized zirconia (YSZ) ceramic particles as a surrogate for UO_2. Particle ALD is performed by sequentially dosing WF_6 and Si_2H_6 into a fluidized bed of YSZ powder, which deposits a tungsten metal coating by self-limited surface reactions to create a nearly conformal 350 nm coating (Figure 1). Thermal treatments are used to evaluate the stability of the coated powders in a dilute hydrogen environment up to 1773 K. Further characterization of the powder by SEM-EDS after high-temperature hydrogen exposure is used to elucidate microstructural changes to the coating and fuel surrogate. To evaluate the ALD tungsten coating under more realistic NTP conditions, a test matrix is developed to compare the performance of the W-coated YSZ particles to the uncoated baseline YSZ particles in a variety of refractory metal matrices using spark-plasma sintering for consolidation. The consolidated cermet samples are then treated with pure hydrogen at >2000 K in MSFC's CFEET facility to simulate conditions approximating NTP engine operation. Microstructural analysis with SEM and XRD, as well as characterization of surrogate fuel mass loss is used to evaluate the effectiveness of the ALD tungsten coating in preventing cermet degradation. This is the first instance of an ALD coating being incorporated into an NTP cermet via SPS, as well as the first test of such a coating under realistic operating conditions. This research advances our understanding of how refractory metal ALD coatings behave under high-temperature and introduces a new technique to improve the durability of NTP engines for deep-space exploration.

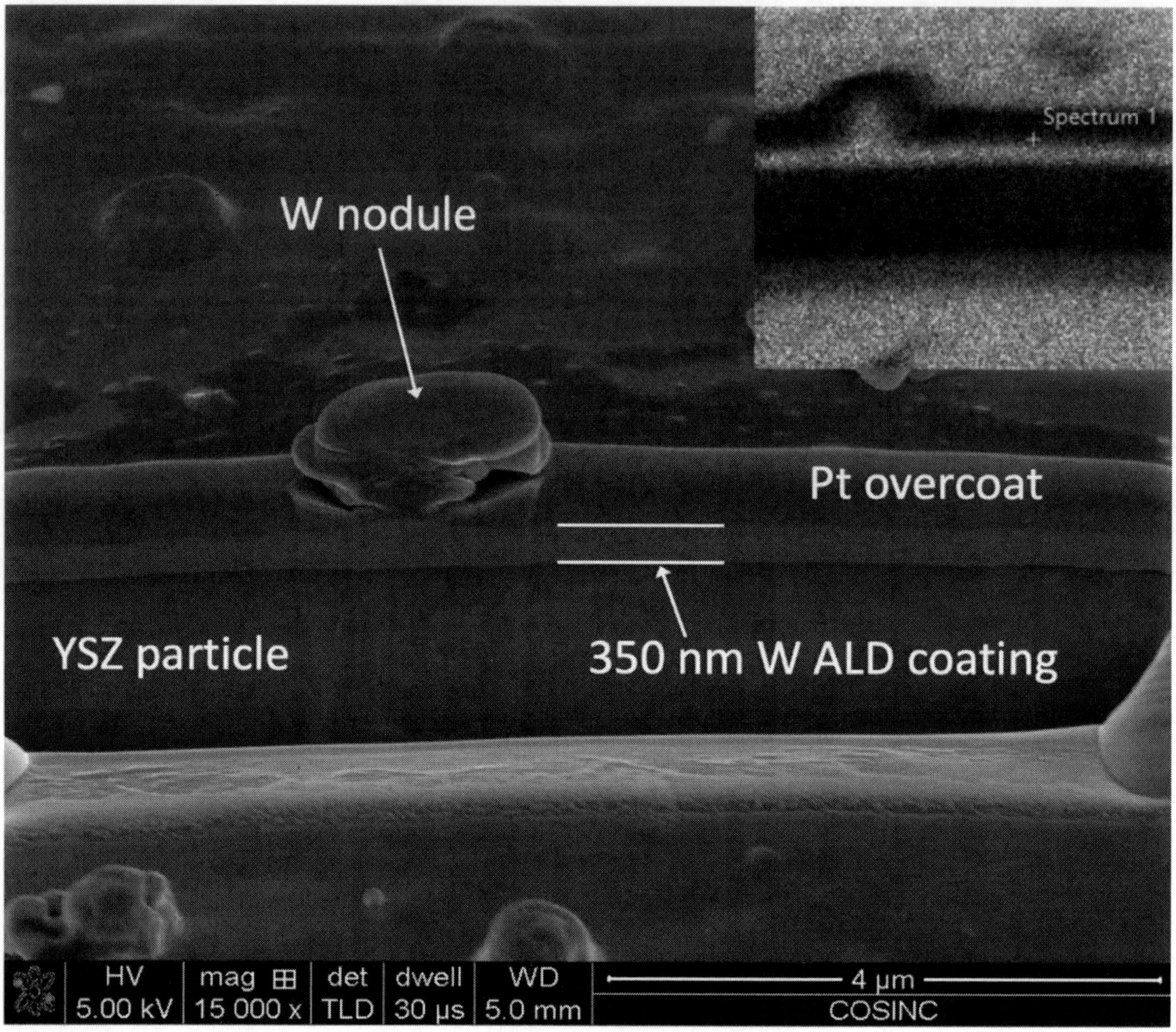

Figure 1: Cross-sectional SEM image of the ALD W coating exposed by FIB milling. Inset: EDS map of W Mα1 emission.

536o Designing Targeted Magnetic Nanoparticles Via Click Chemistry to Diagnose Inflammatory Bowel Disease

Shno Asad, Christel A.S. Bergström, Alexandra Teleki

Introduction: Superparamagnetic iron oxide nanoparticles (SPION) are promising for targeted drug delivery. They can be easily functionalized with targeting ligands, exhibit hyperthermia for triggered drug release and are active as MRI contrast agents.[1] However, due to the lack of scalable and reproducible synthesis methods, few SPION-based medicinal products have reached the clinics. Flame spray pyrolysis (FSP) is an industrially established synthesis process that allows large-scale manufacture of SPIONs. We have developed a single-step flame process for encapsulating SPION with a thin SiO_2 coating[2,3] that increases their biocompatibility and facilitates further functionalization with e.g. targeting ligands by click chemistry. We hypothesize that by functionalizing SPIONs with ligands that target overexpressed biomarkers in the intestinal tissue, we can develop MRI active biosensors for monitoring inflammatory bowel disease (IBD) progression *in vivo*.

Methods: Pure and manganese-doped ($Mn_xFe_{3-x}O_4$; x=0.25, 0.5, 0.75) iron oxide nanoparticles with a thin SiO_2-coating were made through flame spray pyrolysis.[1] T2 relaxivity of synthesized SPIONs was measured on a preclinical MRI system of 3T magnetic field strength. SPIONs were modified by installing an initial amine functional group through silylation reaction using 3(aminopropyl)triethoxysilane (APTES) and analyzed using Fourier Transform Infrared (FTIR) spectroscopy. Thermogravimetric analysis (TGA) was used to compare the amount of organic material on the SPION surface between different functionalization strategies. An organic linker with a terminal alkyne was synthesized through EDC-NHS cross coupling, which was confirmed by nuclear magnetic resonance (NMR). The hydrodynamic particle size was measured by dynamic light scattering (DLS) and the zeta potential by electrophoretic light scattering (ELS). SPION surface is conjugated with IBD targeting ligand via click chemistry by using SiteClick™ Antibody Azido Modification Kit. Targeting capability of the functionalized SPIONs are evaluated *in vivo* in a mouse model for acute colitis where inflammation in the distal colon is induced by addition of dextran sulfate sodium in their drinking water.[4]

Results: MRI measurements demonstrated that manganese-doped SPION have moderately higher T2 relaxivity, with Mn0.5-doped particles exhibiting the highest. Successful amine-functionalization of SPION using APTES was confirmed by FTIR spectroscopy. Hydrodynamic size did not change before and after APTES-modification, however, the zeta potential changed considerably from ζ = -29,3 mV for SPION to ζ = 28,7 mV for the APTES-functionalized SPION. The linker attached on the SPION surface was obtained through EDC/NHS cross coupling reaction of 6-heptynoic acid (94 % yield), confirmed by NMR spectroscopy. Global proteomics has been used to identify ICAM-1 as a promising target luminal biomarker in *in vitro* and *in vivo* preclinical models for IBD.[4] In our ongoing work, we are coupling the linker-modified SPIONs with antibodies for ICAM-1 and evaluating their targeting capability after oral administration to our *in vivo* model for IBD.

Conclusions: Local diseases in the gastrointestinal tract (GIT), such as IBD, can be targeted using functionalized iron oxide nanoparticles. The magnetic properties of the particles makes them useful for the development of an MRI-based diagnostic platform to detect and localize diseases in the GIT.

References: [1] Ling, D., Hyeon, T., *Small.* **9**, 9-10, (2013). [2] Teleki, A., M. Suter, P.R. Kidambi, O. Ergeneman, F. Krumeich, B.J. Nelson, and S.E. Pratsinis, *Chem. Mater.* **21**, 10, (2009). [3] Teleki, A., F.L. Haufe, A.M. Hirt, S.E. Pratsinis, and G.A. Sotiriou, *RSC Adv.* **6**, 26, (2016). [4] Asad, S., Wegler, C., D. Ahl, C.A.S. Bergström, M. Phillipson, P. Artursson and A. Teleki, *J. Pharm. Sci.* **110**, 1 (2021).

409b Packing Dynamics of Polymer Powders at High Temperature

Salvatore Pillitteri, Filip Francqui, Aurelien Neveu, Geoffroy Lumay

Powder characterization is of huge importance for process improvement. As an example, for Additive Manufacturing, methods based on selective layer melting preheat the powder bed higher than the ambient temperature to decrease the amount of energy needed for the laser irradiation to melt the powder. This is especially the case for polymer powders [1]. However, while environmental conditions are known to influence the powder characterization, measurements are rarely performed at constant humidity and temperature or at the same environmental conditions that the process undergone by the powder. This can lead to inconsistent characterization of the powder behaviours due to seasonal change or high differences between measurement and process conditions. For this reason, powder characterization at controlled environmental conditions would be relevant for process improvement [2].

Packing is a popular method to characterize powder density and flowability. In this work, we investigated how the temperature can influence the packing dynamics of plastic powders. We performed tap density measurement with the GranuPack high temperature. The experimental protocol is the following: a steel tube, in which is poured the powder, performs a succession of free falls of 1mm of height, called "taps". After each tap, the height of the granular pile is measured by a laser and the bulk density is computed from the knowledge of the tube diameter and the mass of the powder. The bulk density curve as a function of the taps number is then obtained. The temperature of the tube is controlled by a surrounding electrical heating jacket which can heat from ambient temperature to 200°C.

Significative changes between bulk density curves are observed when temperature increases, affecting both the packing range and the packing dynamics. Indeed, we observed that the initial and tapped density after 500 taps, defined as ρ_0 and ρ_{500} , decrease with the temperature. The packing kinematics is generally described by the Hausner ratio $Hr=\rho_{500}/\rho_0$. We also observe a decrease of Hr as the temperature increases. These results indicate an evolution of particle properties and highlight the importance of temperature for powder characterization.

References

[1] Lumay, G., Francqui, F., Detrembleur, C. and Vandewalle, N., 2020. Influence of temperature on the packing dynamics of polymer powders. Advanced Powder Technology, 31, 4428-4435.

[2] Neveu, A., Francqui, F. and Lumay, G., 2021. Packing dynamics of powders at high temperature. In EPJ Web of Conferences, 249, 12001.

409c A Revolutionary, Evolutionary Approach to the Characterisation of Particulate Media: From Bulk Measurements to Micromechanical Properties

Kit Windows-Yule

A revolutionary, Evolutionary Approach to the Characterisation of Particulate Media: From Bulk Measurements to Micromechanical Properties

There exists a wide array of commercial powder characterisation tools capable of measuring a diverse assortment of bulk powder properties, such as flowability or cohesive index. Many of these properties are, however, purely heuristic and offer little physical insight into the micromechanical properties and behaviours of the materials characterised. This issue is troublesome not only in that it inhibits our understanding of particulate systems, but also means that numerical models thereof (e.g. discrete element method simulations) which require detailed knowledge of powders' microscopic (particle-level) properties cannot be calibrated using commercially-available devices. While in academia researchers have created custom systems for the measurement of particle properties, these ad hoc approaches are not well-validated, nor are they viable for industry.

A lack of knowledge regarding the micromechanical properties of particulate media not only inhibits our fundamental understanding thereof, but also prevents the meaningful application of numerical simulations such as the discrete element method (DEM), which may otherwise prove transformative to a variety of particle-handling industries.

In this talk, we present a novel approach to powder characterisation, using evolutionary algorithms coupled to DEM digital twins [link] of widely-used powder characterisation devices to autonomously determine the microscopic properties of powders from conventional bulk measurements made using said tools. The tool, Autonomous Characterisation and Calibration using Evolutionary Simulation (ACCES) [link] operates, in broad terms, as follows:

(i) A bulk measurement of the material of interest is performed using one or more chosen powder characterisation devices. This may be, for example, the free surface shape formed by a dynamic angle of repose tester (see figure).

(ii) A simulation of said characterisation device(s) is set up.

(iii) Known particle and system properties are set as '*fixed*', unknown properties as '*variable*'.

(iv) An initial '*population*' of simulations is set up by applying random '*mutations*' to *variable*

(v) DEM simulations are run for each '*phenotype*' (i.e. parameter set) created, modelling the real-life procedure of the characterisation device in full.

(vi) The '*fitness*' of each phenotype is assessed based on a given '*cost function*', i.e. its ability to reproduce the experimental measurement(s). This may be, for example, the integral difference between an experimentally-measured and DEM-simulated free surface (see figure).

(vii) The fittest phenotypes are further mutated, to produce a new '*generation*' of simulations.

(viii) Steps (v) to (vii) are repeated until a suitable parameter set, representing the true values of the particles' micromechanical properties, is found.

Unlike conventional methods for the determination of particle-level properties from bulk measurements, which are highly labour-intensive, the ACCES process is entirely autonomous, thus removing arguably the most significant bottleneck from the characterisation process.

As well as acting as a micromechanical characterisation tool, the same methodology may also be applied *directly to a piece of industrial process equipment*, allowing the fully-autonomous calibration of a DEM simulation thereof.

In this talk, we provide an overview of the ACCES methodology, alongside examples of its application to real industrial problems and materials.

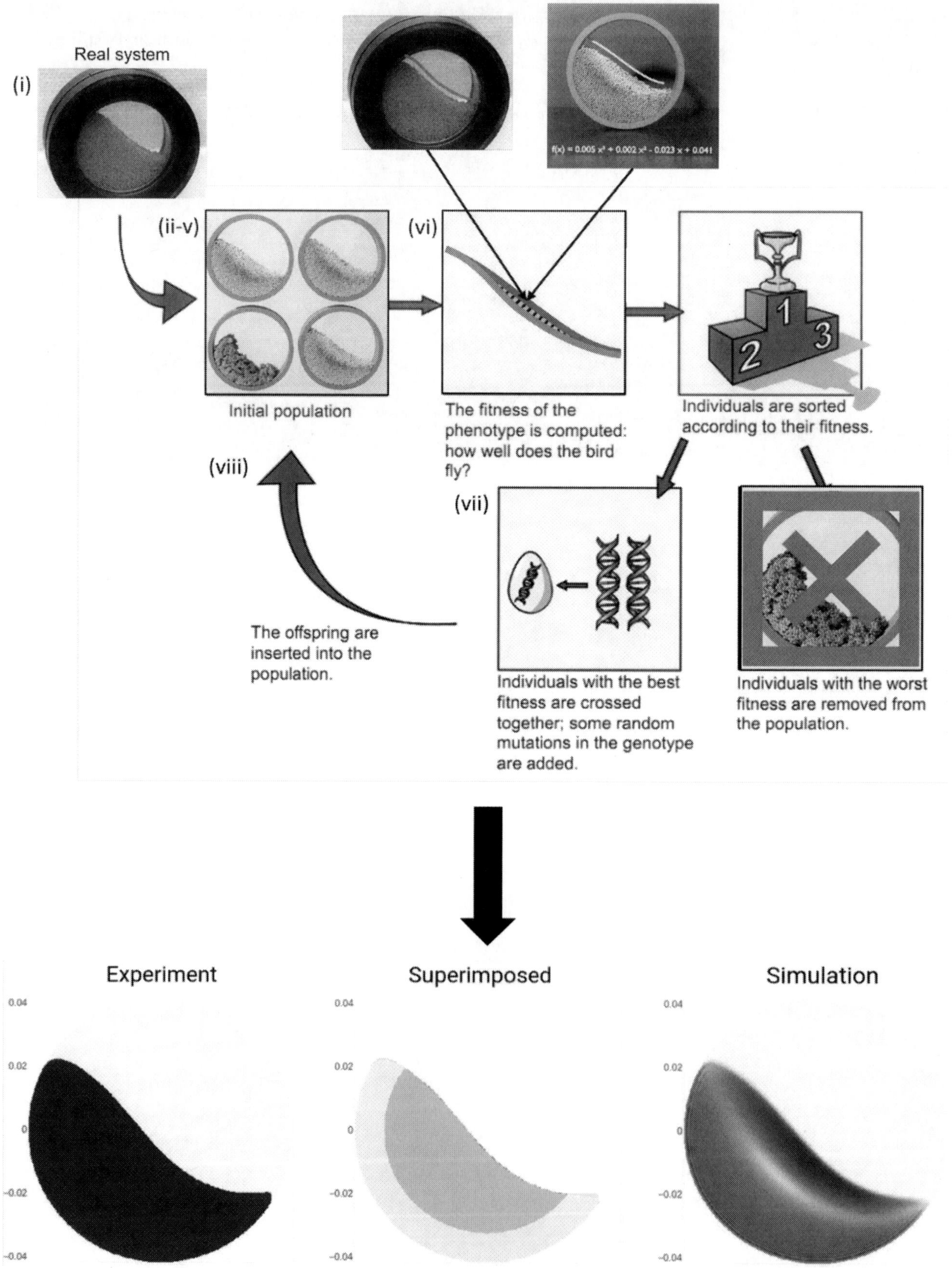
Real system
(i)
f(x) = 0.005 x³ + 0.002 x² - 0.023 x + 0.041
(ii-v)
(vi)
Initial population
The fitness of the phenotype is computed: how well does the bird fly?
Individuals are sorted according to their fitness.
(viii)
(vii)
The offspring are inserted into the population.
Individuals with the best fitness are crossed together; some random mutations in the genotype are added.
Individuals with the worst fitness are removed from the population.
Experiment
Superimposed
Simulation
0.04
0.02
0
-0.02
-0.04

409d Shear-Induced Dilation of Constrained Consolidated Powder Beds

Tony Thornton, Katrina Brockbank, Jamie Clayton, Amalia Thomas

Dilatancy, the phenomenon by which compacted dense granular materials, including powders expand when sheared was first observed by Reynold [1] in 1885. This phenomenon is attributed to the formation of stress chains, consisting of interlocking particles, which rotate during shear, resulting in the expansion of the powder bed [2]. Whilst this behaviour is well recognised, the bulk of scientific investigation on dilatancy has been conducted from a viewpoint of soil (and sand) mechanics rather than powders, as such these consider dilatancy at significantly higher consolidation loads then those routinely encountering during powder handling [3, 4].

The limited studies on powder bed dilatancy have primary used Discrete Element Method (DEM) models or similar [5-7]. Based on these studies, alongside some experimental work, it is generally accepted the particle size will affect the degree of bed expansion during shear, with larger particles resulting in increased dilation [8-10].

Other studies have focused on the impact of inter-particular friction on dilatancy, as this will influence the ability of the powder to from stable stress chains. Again, there is general agreement that increasing inter-particular friction will increase bed dilation [5, 6, 11].

Finally, the impact of Applied Normal Load (ANL) has also been investigated, however the results are conflicting, with several studies showing that the ANL has little to no effect, while other studies have demonstrated a link between Applied Load and dilatancy [8, 12].

Interestingly, the majority of these studies have not considered constrained systems where there is limited or no free volume for the powder to expand into. Therefore, the main aim of this study was to investigate the impact of the ANL on powder bed dilation in constrained systems. Further to this, the impact of particle size and powder flow properties were also investigated.

Using a modified shear cell test, Calcium Carbonate samples (Eskal 500 to Eskal 150) were critically consolidated at ANL values ranging from 0.25kPa up to 20kPa. The ANL was then reduced to 0.01kPa and the powder bed sheared at constant volume. The resultant Generated Normal Stress (GNS) and Shear Stress (SS) values were recorded, alongside changes in the bulk density. Flow properties (Dynamic Flow, Aeration and Shear Cell analysis) and bulk properties (Permeability and Compressibility) were also evaluated for these samples using an FT4 Powder Rheometer® (Freeman Technology, Gloucestershire, UK).

The results demonstrate that the GNS, resulting from dilation of a fixed volume bed, correlates with bulk density changes, rather than a direct correlation with ANL. This would explain the conflicting results for previous studies as bulk density changes are also related to ANL, as further demonstrated by the Compressibility results. At lower ANL, the powder bed exhibited larger relative changes in bulk density and subsequently a greater degree of dilation. As ANL increases, the relative change in bulk density decreases and eventually plateaus, with a similar trend observed in GNS values.

Interestingly, the ratio of GNS to ANL is substantially higher at lower ANL values, with GNS even exceeding ANL in some cases. This was most notable for the larger grades of Eskal, which is consistent with the literature. These results also suggest that consideration of powder bed dilation is required when undertaking Shear Cell analysis at low Applied Normal Load as dilation will have a greater influence on achieving and maintaining the target ANL, especially due to the rapid increase in stress during shear, and will also be variable. In terms of measuring Shear Stress at lower ANL, good correlation was observed between the GNS and SS values, with most Eskal grades exhibiting either a linear or 2nd order polynomial relationship.

As discussed, GNS typically increased with particle size, however, the smallest grade of calcium carbonate, Eskal 500 (D_{50} 0.5μm), deviated from this trend, presenting higher than expected GNS values. This sample was far more compressible than the other grades, exhibiting greater changes in bulk density. Whilst this may explain the larger GNS values at higher ANL, higher compressibility values are typically associated with less efficient particle packing and entrained air. A looser packed powder bed might be expected to accommodate displaced particles more easily in the shear zone, thereby reducing dilation and GNS. Comparison of Specific Energy (SE), a measure of mechanical interlocking and inter-particular friction [13], was notably higher for this sample. Results from previous studies [5, 6, 11] suggest that increased inter-particular friction and mechanical interlocking likely contributed towards the higher than expected GNS values for this sample. As such dilatancy can still be problematic when handling powders of smaller particle size.

The use of a modified shear cell was demonstrated to be a suitable approach for investigating powder dilation in constrained systems. Using this technique, it was possible to investigate the relationship between ANL, bulk density and dilation. The results reinforced the results of previous DEM studies, demonstrating that both particle size and inter-particular friction can influence powder bed dilation.

1. Reynolds, O., *LVII. On the dilatancy of media composed of rigid particles in contact. With experimental illustrations.* The London, Edinburgh, and Dublin Philosophical Magazine and Journal of Science, 1885. **20**(127): p. 469-481.
2. Schulze, D., *Powders and Bulk Solids: Behavior, Characterization, Storage and Flow.* 2021: Springer International Publishing.
3. Tsegaye, A.B., *Cyclic stress-dilatancy relations and associated flow for soils based on hypothesis of complementarity of stress-dilatancy conjugates.* arXiv preprint arXiv:2101.04601, 2021.
4. Szypcio, Z., *Stress-Dilatancy for Soils. Part I: The Frictional State Theory.* Studia Geotechnica et Mechanica, 2017. **38**(4): p. 51-57.
5. Shi, H., et al., *Steady state rheology of homogeneous and inhomogeneous cohesive granular materials.* Granular matter, 2020. **22**(1): p. 1-20.
6. Yang, Z.X., J. Yang, and L.Z. Wang, *On the influence of inter-particle friction and dilatancy in granular materials: a numerical analysis.* Granular Matter, 2012. **14**(3): p. 433-447.
7. Amirpour Harehdasht, S., et al., *Influence of particle size and gradation on shear strength–dilation relation of granular materials.* Canadian Geotechnical Journal, 2019. **56**(2): p. 208-227.
8. Louati, H., et al., *Qualitative and quantitative DEM analysis of cohesive granular material behaviour in FT4 shear tester.* Chemical Engineering Research and Design, 2019. **148**: p. 155-163.
9. Faqih, A., et al., *Flow - induced dilation of cohesive granular materials.* AIChE Journal, 2006. **52**(12): p. 4124-4132.
10. Obregón, L., A. Realpe, and C. Velázquez, *Mixing of granular materials, part II: effect of particle size under periodic shear.* Powder Technology, 2010. **201**(3): p. 193-200.
11. Babu, V., et al., *Dilatancy, shear jamming, and a generalized jamming phase diagram of frictionless sphere packings.* Soft Matter, 2021. **17**(11): p. 3121-3127.
12. Mofiz, S.A., M.R. Taha, and D.C. Sharker. *Mechanical stress-strain characteristics and model behavior of geosynthetic reinforced soil composites.* in *17th ASCE Eng. Mechanics Conf.* 2004.
13. Freeman, R., *Measuring the flow properties of consolidated, conditioned and aerated powders — A comparative study using a powder rheometer and a rotational shear cell.* Powder Technology, 2007. **174**(1–2): p. 25-33.

409f Flowability Measurements and Rheological Investigations of Volcaniclastic Debris Flows from Campania Plain (southern Italy)

Ilaria Rucco, Chongqiang Zhu, Fabio Dioguardi, Lyes Ait Ali Yahia, Lizeth Caballero, Mauro A. Di Vito, Damiano Sarocchi, Raffaella Ocone

In perivolcanic areas such as the Campania Plain (southern Italy) is not so uncommon the remobilisation of the pyroclastic deposits emplaced on the slopes of the volcanoes and of the surrounding reliefs, giving rise to very hazardous phenomena known as volcaniclastic debris flows (Di Vito et al., 2019). They are very unpredictable phenomena that can happen during or soon after an eruption, but also tens or hundreds of years later, representing both a short and long-term hazard factors for people, buildings and natural environment (Zanchetta et al, 2004). The main triggering mechanism is represented by intense and/or prolonged rainfalls (Sulpizio et al., 2006 and references therein) that are becoming more frequent as a consequence of climate change.

The novelty of this project is combining engineering and geology to analyse samples of volcaniclastic debris flows related to the AD 472 Pollena eruption from Somma-Vesuvius volcano, in Campania region, to acquire fundamental knowledge that will help develop constitutive equations to use in existing numerical models. Since very few rheological analysis with real samples are available, the final aim is to develop a tool that can be applied to all the volcaniclastic debris flows and that can be used to manage the risks associated with these poorly known phenomena.

The experiments have been carried out with the FT4 powder rheometer (Freeman Technology), commonly used in the chemical and process industries, that has been modified for our purposes.

Shear cell and wall friction tests give information on the flowability of the materials through charactering and measuring relevant parameters such as the angle of internal friction and the wall friction angle. Compressibility tests show the change in density and volume as a function of the applied normal stress. The relationship between shear stress and shear rate is also quantified using a 3D printed attachment in a Couette flow configuration (Ait Ali Yahia et al., 2020).

Preliminary results show that the particles' size influences the compressibility but not so much the shear stress; the wall friction angle depends on the roughness of the wall friction head and also on the grainsize distribution. Concerning the rheological analysis, the polydisperse samples and the finest grain sizes are difficult to analyse, and we are currently working on devising the correct measuring protocol and we are considering working with the suspensions. Future work will see the development of a systematic experiments' campaign in Mexico, where different rheometers will be used and large-scale experiments will be performed, in order to link the rheology measurements in the laboratory to the dynamics observed in the large-scale experiments.

References:

Ait Ali Yahia L., Piepke T.M., Barrett R., Ozel A., Ocone R. (2020). Development of a virtual Couette rheometer for aerated granular material. AIChE J., e16945.

Di Vito M.A., Talamo P., de Vita S., Rucco I., Cesarano M. (2019). Dynamics and effects of the Vesuvius Pomici di Avellino Plinian eruption and related phenomena on the Bronze Age landscape of Campania region (Southern Italy). Quaternary International 499, 231-244.

Sulpizio R., Zanchetta G., Demi F., Di Vito M.A., Pareschi M.T., Santacroce R. (2006) - The

Holocene syneruptive volcaniclastic debris flows in the Vesuvian area: Geological data as a guide for hazard assessment. Geol. Soc. Am. Spec. Pap., 402, 203-221.

Zanchetta G., Sulpizio R., Pareschi M.T., Leoni F.M., Santacroce R. (2004). Characteristics of May 5-6, 1998 volcaniclastic debris flows in the Sarno areas (Campania, southern Italy): relationships to structural damage and hazard zonation. Journal of Volcanology and Geothermal Research 133, 377-393.

206a Impregnation of Catalyst: Batch and Continuum Processes

Maria Tomassone, William Borghard

This project focuses on developing science-based methods for designing, and optimizing the catalyst impregnation process in batch and continuous equipment. Discrete element method (DEM) simulations are combined with parametrically and geometrically identical experiments to validate simulations. The DEM model is coupled with a novel algorithm that allows the transfer of fluid (water with metal), viscous and non- viscous, to and between particles. In previous meetings, the simulation with the algorithm has been compared and validated with matching experiments. Good agreement has been demonstrated in the results for the relative standard deviation (RSD) of mixing and liquid distribution for both double cone blender and cylindrical vessels. Using the validated DEM model, we have also investigated the effects of rotation speed, spray pattern, support size andamp; morphology, and addition of baffles on the overall mixing. We investigated the effect of baffles and elucidate the relevant factors that control a continuous impregnation process. The two main objectives for this work are: 1) obtain fundamental understanding of continuous impregnation process for particles and powders and 2) investigate process parameters and develop scientific relationships to increase the efficiency of continuous impregnation. Some key parameters for the continuous impregnation process include the residence time distribution, the axial dispersion coefficient, the product RSD and the mass transfer coefficients. We have done a systematic study covering before and after steady state was reached focusing on the RSD and water content of tracers particles in the collection box. The focus was to reveal details about the parameters that affect the homogeneity of the particle bed, such as the mean residence time (MRT) and the residence time distribution (RTD) of the tracer particles, the separation of the nozzles and the flow rate distribution in the nozzles. We studied 3 different RPMs (1, 3 and 5), and 3 different angles (1, 3 and 5) in two different flow rate configurations (even and uneven) with 4 nozzles. We observe very interesting results. Our work shows that the bed appears to contain two regimes: Towards the beginning of the vessel: neighboring particles tend to remain close together (Regime 1). As tracer particles move through vessel, they tend to mix. Towards the end of the vessel, particles that were close together in the beginning of the vessel are now mixed (Regime 2). The two Regimes in the vessel should be treated differently.More water should be sprayed in Regime 1, where neighboring particles remain close together. Particles will be exposed to similar amounts of water and have more time to exchange fluid. Less water should be sprayed in Regime 2: water content uniformity will be benefited by water exchange between particles. We obtained better results when 70% of the total amount of fluid is sprayed in Regime 1 and 30% in Regime 2.

In summary, we developed a systematic approach to study the continuous impregnation process. We are now in a good position to test and optimize the effect of various parameters, the number of nozzles, the nozzle spacing, the wetted area, the drum speed, the Q patterns, etc. During startup, the number of particles inside the cylinder first increases and then reaches a steady state value. The higher the rotational speed, the sooner the steady state is reached and the shallower the bed is. In our studies, we also found that the RSD is proportional to a newly proposed index H, which trends well with RSD for the water content in the collection box and for the tracers after reaching SS, for either even or uneven patterns of flow rates. In all our studies, we see that the effects of inclination angle and rotational speed are consistent with Sullivan's prediction. Additional work is needed to ensure that the continuous process can produce product of a quality equal to or better than the batch process.

206b DEM and Experimental Comparison of Particle Heat Transfer and Geometry Driven Particle Flow Patterns

Jason Schirck, Aaron Morris, Korey Cook, Zhiwen Ma

The discrete element modeling (DEM) technique is used to predict the particle temperature distribution and heat transfer coefficient of an experimental particle heater. The simulation results are then compared to experimental measurements for a small-scale heater. The experimental setup contains rows and columns of tubes with a diamond-like cross-section. The tube surfaces are embedded with heating wires. Figure 1 shows a schematic of the heater (right) as well as an image of the experimental apparatus where particles can be seen flowing around the various heater elements. The heaters are set to 600°C, and the inlet flow of the particles is approximately 400°C to match the experimental conditions. Overall, the particle temperatures between the numerical and the experimental work match within 10%, for several locations within the heater. Also, the heat transfer coefficients predicted by the discrete element simulations is nearly identical to the experimentally determined values. A unique oscillating flow pattern at the heater outlet is observed in the DEM simulations. The oscillatory flow pattern occurs when the upper-most surface of the heater tube is curved.

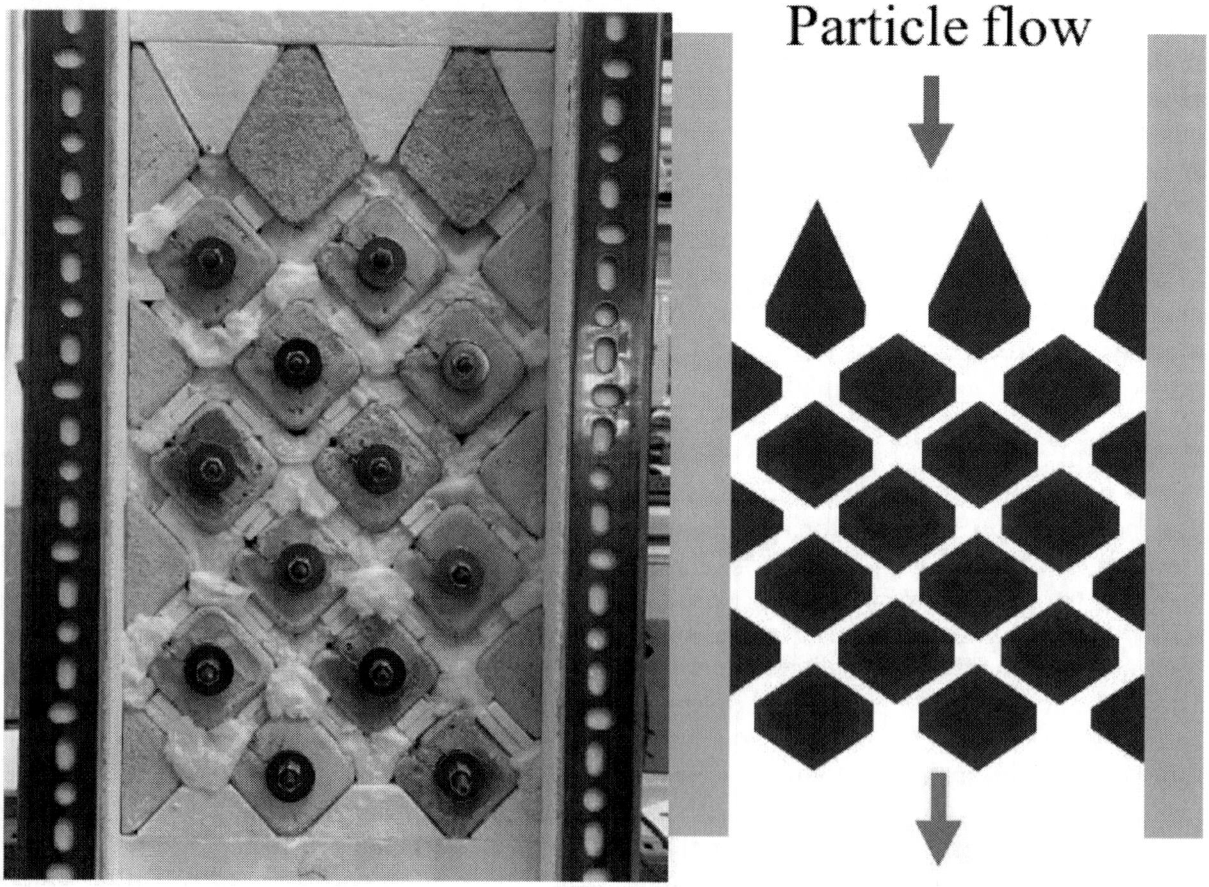

206c Effect of Mechanical Interlocking on Particle-Phase Stress in Sheared Granular Flows

Elizabeth Suehr, Jiecheng Yang, Jennifer Curtis

In order to obtain reliable predictions for the flow behavior of highly angular particles, such as Lunar and Martian soils, mechanical interlocking associated with sustained particle-particle interactions needs to be described. Mechanical interlocking gives rise to large particle-phase stress in granular soils and a significant resistance to deformation. In this study, the effect of particle interlocking in dilute and dense-phase granular flows is investigated via discrete element method (DEM) simulation of a 3D system of 'C'-shaped particles in simple shear flow using Lees-Edwards periodic boundary conditions. The effect of particle shape is assessed by progressively varying the degree of curvature of these 'C'-shaped particles. The particle shape varied from a straight, elongated particle to one in which the straight particle is folded in half. For each particle shape, a total normalized interlocking time and distribution of particle interlocks are determined over a range of solid concentrations. At low solid volume fractions, the normal and shear stresses are proportional to the inverse of the projected area of the particle in the plane perpendicular to the flow, consistent with the stress behavior of elongated particles [1]. For larger solid volume fractions, normal and shear stresses increase with increasing interlocking, showing a direct dependence between stress and degree of mechanical interlocking.

[1] Y. Guo, C. Wassgren, W. Ketterhagen, B. Hancock, B. James and J. Curtis, "A Numerical Study of Granular Shear Flows of Rod-like Particles Using the Discrete Element Method", *Journal of Fluid Mechanics*, **713**, 1-26 (2012)

206d Continuum Modeling of Cohesive Powder Flow

Sameer Kamath, Sean Garner, Marcial Gonzalez, William Ketterhagen, Carl R. Wassgren

This presentation will describe a continuum model for predicting the flow of a cohesive powder used in pharmaceutical manufacturing. The model consists of an Eulerian-Lagrangian finite element method simulation using a modified Drucker-Prager Cap (mDPC) material model implemented in the commercially-available Abaqus Explicit software package. The existing issues of excessive dilation with the Abaqus built-in DPC model have been improved by modifying the plastic potential surface in the shear zone. Material properties are obtained from the literature as well as laboratory tests. Simulations of powder flow in a feed screw are performed and compared to experimental observations.

206f DEM Study of a Vibrational Powder Transport System

Martina Trogrlic, MSc, Dalibor Jajcevic, Johannes G. Khinast, Pankaj Doshi, Barry Ager, Tata Venkata, Stephen Franklin, David Barling

Vibrational transport systems are used in various industries to aid the transport of the granular material between unit operations. The system tipically consists of one or several parts that oscillate with the help of a vibration source, which is controlled remotely. The parameters that control the vibration are the amplitude and the frequency of the vibration, and the length of the vibration period, which can be adjusted.

The horizontally vibrating system used in this work is shown in Fig. 1. It is essentially a hopper attached to an elbow-shaped tube, with a diameter size reduction in the horizontal part. Powder is introduced from the top in the filling hopper and transported downstream to the next unit operation. The vibration source is fixed to the horizontal transport tube, and the whole system oscillates as one part.

The challenges, in this case, come from the particle agglomeration, system clogging, and flow blockage in the hopper and transport tube, especially because the transported material has poor flowability and is highly cohesive. Experimental trials using three powders with different cohesivity showed that the operating frequency affects the flowability of the powder.

To better understand why the operating vibrational frequency affects the particle flow behavior, the Discrete Element Method (DEM) was used to simulate the process [1], [2]. With the DEM approach, it is possible to investigate various process conditions and equipment designs in a short period, gaining process insight that is difficult or even impossible to get experimentally.

The vibrational transport trials were closely replicated in the DEM model, with the same operating conditions and equipment design. The three powders were calibrated prior to the investigation (DEM reproduction of the FT4 shear cell and compression tests [3]), in order to closely replicate the behavior of the real powders. The particle size was distributed with a normal distribution function and scaled in comparison to the original one in order to speed up the simulations.

The DEM results revealed an interesting mechanism, i.e., the segregation of the particles with small diameter in several locations in the system, including the fill hopper and the horizontal tube. The investigation of the particle stress distribution in the system showed a peak of the normal and shear stress at the position of the segregated particles. The fact that these positions overlap led to the conclusion that the system clogging is caused by the agglomeration of the small particles.

The DEM investigation showed that the frequency change affects the position and trajectories of the small particles. The small particles are dispersed more in the system when the operating frequency is lower, which reduces the segregation and helps to improve the powder flowability. Our finds were confirmed by the experiments as well.

[1] D. Jajcevic, E. Siegmann, C. Radeke, and J. G. Khinast, “Large-scale CFD-DEM simulations of fluidized granular systems,” *Chem. Eng. Sci.*, vol. 98, pp. 298–310, 2013, doi: 10.1016/j.ces.2013.05.014.

[2] E. Siegmann, D. Jajcevic, C. Radeke, D. Strube, K. Friedrich, and J. G. Khinast, “Efficient Discrete Element Method Simulation Strategy for Analyzing Large-Scale Agitated Powder Mixers,” *Chemie-Ingenieur-Technik*, vol. 89, no. 8, pp. 995–1005, 2017, doi: 10.1002/cite.201700004.

[3] P. Toson *et al.*, “Continuous mixing technology: Validation of a DEM model,” *Int. J. Pharm.*, vol. 608, no. September, p. 121065, 2021, doi: 10.1016/j.ijpharm.2021.121065.

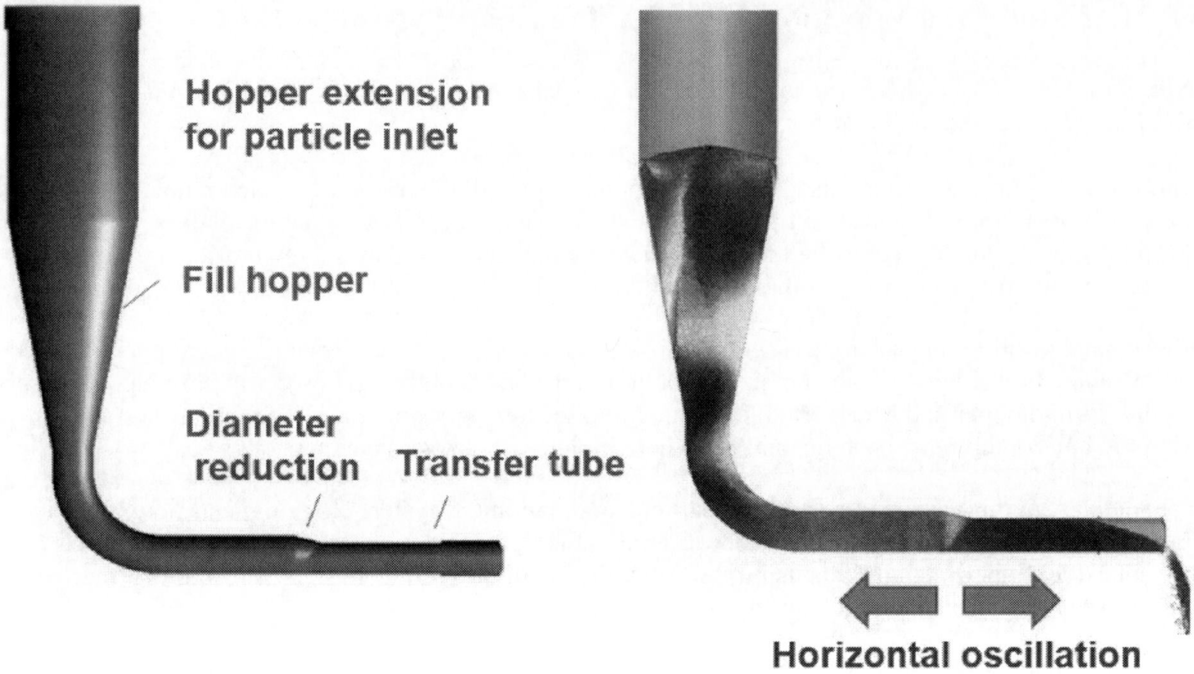
Hopper extension
for particle inlet
Fill hopper
Diameter
reduction
Transfer tube
Horizontal oscillation

206g Multiscale Computational Fluid Dynamics Method for Slug Flow Reactor Simulation

Shin Hyuk Kim, Moo Sun Hong, Jay H. Lee, Richard D. Braatz

Millifluidic slug-flow reactors recently have been attracting attention for the manufacturing of micron-sized particles. However, simulation methods for predicting the physicochemical phenomena inside the slug-flow reactor have been lacking. Microparticle formed within slugs spontaneously formed in gas-liquid multi-phase flows induces gas-liquid-solid three-phase flows that interact to result in the final particle size distribution at the outlet of the reactor. A multi-scale simulation method that can predict three-phase flows and the resulting particle size distribution in slug-flow reactors is desired, for use in process design and control.

This work introduces a MultiPhase Particle-In-Cell model coupled with Volume of Fluid and Population Balance Equation (MP-PIC-VPBE), which is an upgraded version of MP-PIC-PBE [1, 2]. The MP-PIC-VPBE method predicts the slug flow formation in the gas-liquid phase based on the VOF method and simultaneously calculates the particulate flow pattern and particle size change in the liquid-solid phase using Lagrangian parcels combined with PBE. In addition, the energy equation, species equation, and equation of state are implemented to accurately predict the effects of temperature, concentration, and density of the gas-liquid phase on PBE kernels about the particle size distribution.

In this study, the newly developed method is used to simulate the crystallization of L-asparagine monohydrate (LAM) in a slug-flow reactor and this model is validated through direct comparison with experimental data. An optical microscope combined with a digital camera and a transparent tubular reactor is used to observe slug flow and particulate flow patterns inside the reactor, in a configuration similar to past work [3]. Also, the particle distribution of the end product is measured using a high-magnification stereo microscope and compared with the simulation results. The MP-PIC-VPBE is able to simulate effects of changing process operations that are not predicted well by simpler models.

[1] S. H. Kim, Richard D. Braatz, and Jay H. Lee, "Multi-scale fluid dynamics simulation based on MP-PIC-PBE method for PMMA suspension polymerization," *Computers and Chemical Engineering*, 152, 107391, 2021.

[2] S. H. Kim, Richard D. Braatz, and Jay H. Lee, "Multi-phase particle-in-cell coupled with population balance equation (MP-PIC-PBE) method for multiscale computational fluid dynamics simulation," *Computers and Chemical Engineering*, 134, 106686, 2020.

[3] Mo Jiang and Richard D. Braatz, "Low-cost noninvasive real-time imaging for tubular continuous-flow crystallization," *Chemical Engineering andamp; Technology*, 41(1), 143-148, 2018.

206h Effect of Particle Shape on Deposition Behavior in Cascade Impactor

Ryosuke Mitani, Shuji Ohsaki, Hideya Nakamura, Satoru Watano

In human respiratory system, fluid behavior is complicated, and it is difficult for predicting drug particle behavior in human lung. Therefore, cascade impactors are commonly used to assess the size characteristics of aerosols experimentally. However, a lot of questions remain related to the impactor performance, including the influence of particle shape on measured size distributions. Particle shape plays a key role in determining the particle behavior in fluid because fluid drag force varied by the cross-section area of particle.

So far, drag force of irregular shape particle is calculated by the drag coefficient C_D and it was suggested by Ganser [1], Hölzer and Sommerfeld [2], Bagheri and Bonadonna [3] and so on. However, these models were applied to the straight flow and not to the complicated flow such as the cascade impactor. Therefore, irregular particles behavior for inhalation was still unknown.

To analyze the irregular particles behavior, computational fluid dynamics and discrete phase model (CFD–DPM) was used to simulate the behavior of both fluid and irregular particles. The fluid drag force of irregular particles was calculated by the suggested models. Needle shaped particles were used as irregular particles for the CFD–DPM numerical simulation. From simulated results, recirculation vortex was observed at upper stage in the cascade impactor. Because of this recirculation vortex, fine particles deposited at upper stage. Also, it was found that the aspect ratio of particles was related to deposition in cascade impactor.

References

[1] G. H. Ganser, *Powder Technology,* **77** (1993) 143–152.

[2] A. Hölzer and M. Sommerfeld, *Powder Technol.* **184** (2008) 361–365.

[3] G. Bagheri, C. Bonadonna, *Powder Technol.* **301** (2016) 526–544.

270a Scaling up Segregation Learnings in Industrial Bin Flow

Alexander M. Fry, Jason A. Stamper, Christopher G. Stoltz, John P. Hecht

Recent advancements in continuum and numerical modeling of granular flows provide hope of quantitatively predicting mixing and segregation in powder flows at large scale. We are interested in applying these models to plant-scale bins, and we have studied the scale up of size-driven segregation learnings through bench and pilot scale experiments, alongside continuum and discrete element method (DEM) models. We found that the filling mode of 3D bins is in practice often by intermittent avalanches and that the segregation during bin filling continues to increase inversely with the filling rate well into the intermittent-avalanching regime, down to the lowest flow rates tested. As a result of the intermittent filling mode, the intensity of size segregation did not scale as expected (with the rise velocity of the heap surface) with bin size and fill rate in our experiments. We also used the multiple scale bins in experiment and DEM to design and test bin inserts for altering the discharge flow patterns in the geometrically similar bins, incorporating known design principles from the literature. Open challenges to implementing continuum segregation models in industrial bin flows include linking predictions for the size and frequency of intermittent avalanches during 3D hopper filling to the continuum model for segregation.

270b Drag on a Sphere in Granular Shear Flows

Lu Jing, Julio M. Ottino, Paul B. Umbanhowar, Richard Lueptow

Granular flows consisting of multiple particle species tend to segregate due to differences in particle size, density, or other physical properties. Accurate modeling of the segregation process often requires the knowledge of the effective inter-species drag resulting from particle contacts as different particle species percolate through each other. However, a full picture of the granular drag remains unclear even for the simplest scenario where a single intruder particle migrates in a sheared granular bed. Here, we use discrete element method simulations to explore the drag on a spherical intruder in a uniform-shear granular flow in the absence of gravity. The intruder is displaced by an external force perpendicular to the shear, resulting in an effectively constant intruder velocity (i.e., the effective drag force exerted on the intruder by bed particle contacts counterbalances the applied force). For broadly varied applied forces, intruder properties (size and density ratios), and flow conditions (shear rates and overburden pressures, hence inertial numbers), we find a striking similarity between intruder drag in a granular flow and drag on a sphere in a viscous fluid; that is, the drag coefficient C_d and the intruder Reynolds number Re_i follows a Stokesian form, C_d=8c/Re_i, for Re_i spanning five orders of magnitude, where c lies approximately between 1 and 3, noting that c=3 corresponds to Stokes drag on a sphere in a viscous fluid. Furthermore, an empirical expression is developed to capture the secondary dependence of c on the inertial number and the intruder size and density ratios. Finally, we show that the modified Stokes' drag model leads to a terminal velocity-like formulation for the segregation velocity of intruder particles, matching extensive simulations of gravity-driven segregation in granular flows. This material is based upon work supported by the National Science Foundation under Grant No. CBET-1929265.

270c Flow and Segregation of a Mixture of Passive and Responsive, Magnetically-Driven Rotating Particles

Samuel Wilson-Whitford, Jinghui Gao, William Buckley, David Kramer, James Gilchrist

Gravity-driven granular flows, commonly found in nature in the form of avalanches and landslides and industrial processing, have long piqued the interest of scientists and engineers. Their actuation simplicity and response complexity have designated them as a cornerstone system of study within the soft matter and complex systems communities. Heap formation is a typical behavior of poured granular systems. With continuous addition of a granular material, such as sand, to a flat surface, as a heap forms a continuous gravity driven boundary flow can be observed in which grains are transported from an uphill to a downhill position. The heap itself grows as grains are transferred from the boundary layer to the underlying static heap. The dissipation of potential energy through frictional particle-particle and hydrodynamic particle-fluid interactions within the boundary layer shear flow leads to a fundamental slope angle, known as the angle of repose and the kinematics of the boundary layer itself have been well-documented through granular rheology of these inertial systems.

What has not previously been explored is the behavior of granular systems where shear is introduced to the system through torque applied directly to the particles themselves. In this work, a granular system of 50 micron ferromagnetic Janus particles is prepared. These responsive Janus particles are rotated through the application of magnetic torque by an external rotating magnetic field resulting in uphill granular heaping which for all purposes demonstrates quintessential granular behavior, such as boundary layer flow, but against gravity, giving rise to a negative angle of repose (downhill to uphill). When mixed with other non-functionalized particles, segregation occurs that may rise to well-known fluid instabilities. It is proposed that the behavior is governed frictional control as a direct result of the magnetic field strength. These observations elucidate a whole new way to observe granular systems and have a significant impact on the way in which granular systems could be applied.

270d Experimental Investigation of Segregation in a Rotating Cylinder with Non-Spherical Particles

Sunil Kumar, Anshu Anand

The effect of multiple parameters on mixing two unequal sized non-spherical particles is studied in a rotating cylinder. The result shows the shape of both coarse and fine particle influence mixing. Results are explained with the help of mono-dispersed random packing density and angularity of the particles.

For the coarse particles that are considered, decreasing trend of radial segregation index is as follows: sphere > oblate > prolate > elongated needle. This is strongly correlated to mono-disperse packing fraction for the coarser particles. Figure 1 shows comparison of segregation index for different coarse particle's shapes. For the fine particle, the trend follows sphere > prolate > cube which correlates to the angularity in fine particles.

The extent of segregation is also affected by the initial proportion of fines added to the system. Decreasing the concentration of fine in the initial fill system decreases the extent of segregation. Overall, the segregation in general can be reduced by using non-spherical particles.

270e DEM Study of Non-Spherical Ellipsoidal Particles in a Vertically Vibrated Packed Bed System

Salma Khatoon, Anshu Anand

Uniform mixing is one of the important aspects in many industrial applications such as food, pharmaceutical, chemical, and fertilizer industries. Most of the granular materials found in the industries are non-spherical, among which an important variation from the spherical shape is an ellipsoidal (Oblate and Prolate). In the current study, the Discrete Element Method (DEM) is employed to investigate the mixing behavior of the ellipsoidal particles in the vibrated packed bed system at a fixed amplitude and frequency. The study is carried out with the bidisperse particle by fixing one particle to be a sphere, and varying the aspect ratio (0.25, 0.5, 0.75, 1, 1.5, 2.0, 2.5, 3.0) of another ellipsoidal particle. The velocity fluctuation is observed maximum in the vertical direction. The generation of convection rolls plays an important role in mixing within the vibrated bed. The result shows that the non-sphericity and size of the particle have a significant effect on the mixing behavior. The extent of mixing (SMI) is calculated and observed that non-sphericity improves mixing. The observed simulation result is validated with the experimental result which shows good agreement.

270f A Particle-Force Based Theory for Size Segregation of Binary Granular Mixtures

Anurag Tripathi, Alok Kumar, Mohit Nema, Devang Khakhar

A granular mixture of different size particles flowing over an inclined surface is known to segregate with large particles concentrating near the free surface. In order to understand and predict size segregation of granular mixtures, we first propose a novel method to measure the upward force acting on a single, unconstrained, large particle in a granular medium of small particles flowing over inclined plane using discrete element method (DEM) simulation. Frictional, slightly inelastic, soft deformable spherical particles flowing over an inclined bumpy surface are simulated with few large particles. Four (nine) equidistant intruder particles of large size are sandwiched between two cubic lattice configuration of nontouching small particles at the beginning of the simulations. The particles are allowed to settle under the influence of gravity and the flow is continued until steady state. A harmonic spring force is used to ensure that the intruder particles remain nearly at the same height. In addition, a small repulsive force is used between the intruder particles to prevent them from coming close to each other during the evolution of the flow. Once the flow reaches steady state, the harmonic attraction force and the repulsive force between the intruder particles are switched off. The mass of each intruder particles is set as desired and the simulation is continued. Depending on their mass, the intruder particles may either rise or sink in the flowing layer. The average velocity of the intruder along the vertical direction is obtained as the slope of the average trajectory of many particles by means of a linear fit. Mass of the intruder shows a linear variation with the local viscosity, intruder velocity and diameter, indicating a Stokes' like drag acting on the rising/settling particles. By calculating the mass of the intruder that remains neutrally buoyant in the flowing layer, we measure the net upward force that leads to segregation of large particles near free surface in shear flows. Using this measure of the net force acting on the large size intruder particle, we next formulate a theory to predict the segregation due to size difference. While the predictions agree very well with simulation results across most part of the layer, it fails to capture the behaviour near the free surface where the concentration of large particles is much higher.

We next take into account the fact that the net upward force on a single large particle should reduce as the concentration of the large particle increases and must approach to zero in the limiting case of large particles concentration approaching unity. The measured net upward force on a single large intruder, when corrected for the large particle concentration dependence, yields a segregation flux with cubic dependence on the large particle concentration and linear dependence on shear rate. In addition, a dependence on the inclination angle is also predicted by the theory which is not captured in the empirical segregation flux-based approaches. Accurate predictions of the concentration and the velocity field are done using this approach for steady, fully developed chute flow of binary mixtures for a wide range of compositions and inclinations for two different size ratios. The theoretical formulation suggests a simple, yet quantitative, way of predicting size segregation in size bidisperse mixtures and paves way for predicting segregation of mixtures differing in both size and density.

References:

Anurag Tripathi, Alok Kumar, Mohit Nema and D. V. Khakhar, 'A theory for size segregation in flowing granular mixtures based on computation of forces on a single large particle', Phys. Rev. E., 103 (3), L031301 (2021)

270g Time Dependent Size Segregation of Binary Granular Mixtures Flowing over a Chute

Soniya Kumawat, Vishnu Kumar Sahu, Satyabrata Patro, Anubhav Majumdar, Anurag Tripathi

Transient size segregation and flow behavior of bi-disperse granular mixtures flowing over an inclined surface under the influence of gravity is studied using DEM simulations and theory. A recently developed particle force-based size segregation theory has been shown to predict the steady-state behavior of binary granular mixtures successfully [1]. This promising theory of binary mixtures is explored in detail in this work by comparing the theoretical predictions with DEM simulations. The inter-coupling of size segregation with rheology [2] is accounted to predict various properties of interest such as specie concentration, inertial number, velocity, pressure and shear stress at different times. A two-way coupled continuum model to solve the momentum balance equations along with particle force-based segregation transport equation and mixture rheology is developed using Matlab PDEPE solver. Figure shows the comparison of theoretically predicted time dependent concentration profiles (solid lines) with DEM simulations (symbols) for a binary mixture having equal composition of each species with size ratio $r = 1.25$ flowing over a chute of inclination angle of 25°. The time-dependent evolution of flow properties from the continuum model are found to be in good agreement with the DEM simulations for different size ratios for a wide range of compositions.

References:

[1] Anurag Tripathi, Alok Kumar, Mohit Nema, and D. V. Khakhar. "Theory for size segregation in flowing granular mixtures based on computation of forces on a single large particle". Phys. Rev. E, 103:L031301, 2021.

[2] Anurag Tripathi and D. V. Khakhar. "Density difference-driven segregation in a dense granular flow". J. Fluid Mech., 717:643–669, 2013.

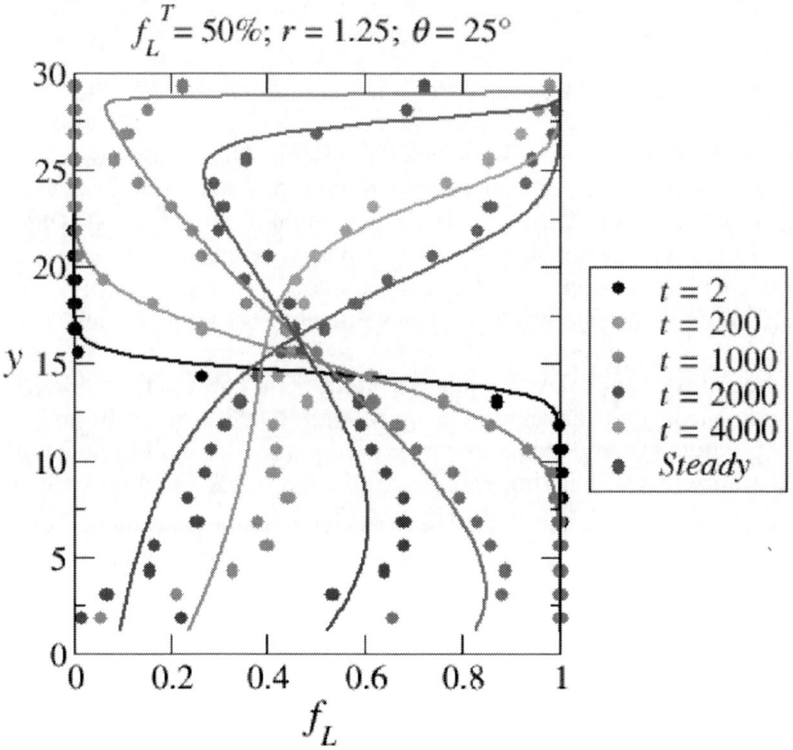

Figure 1. Comparison of theoretical predictions of large size particle concentration (solid lines) with DEM simulations (symbols) for different times.

270h Investigation of Segregation Effect on the Packing Fraction for Density Estimation

Salvatore Pillitteri, Filip Francqui, Aurelien Neveu, Geoffroy Lumay

Granular segregation is a commonly encountered problem during powder handling. Indeed, powders are polydisperse granular materials by nature. The size differences between particles can lead to various effects: small grains can percolate through the porous medium made by larger particles, the Brazil Nuts effect has the tendency to bring the large grains on the top of a granular pile under the influence of vibrations or a powder flow can unmix different species. While homogenous mixing can be time and energy consuming, the next steps of a manufacturing process can segregate the prepared powder mixture during conveying or packing due to inherent vibrations and flow encountered that lead to granular segregation.

Although granular segregation can be observed in manufacturing processes, powder mixtures are often assumed homogeneous for modelization and next steps. However, a segregated powder can lead to unwanted drawback. As an example, for medicines, the powder mixture must have a good homogeneity to guarantee a precise amount of active material and good mechanical properties. Another example is the density of concrete which is highly dependent on the homogeneity of the granular material inside the mixture.

The bulk density of a granular mixture is an important parameter that can influences properties of the final product such as its mechanical properties. We present experimental results on the bulk density of binary mixtures, made of two sizes of glass beads as model material. Different mixtures were prepared with different degrees of homogeneity by forcing the segregation and varying the proportion of small and large particles. For each mixture, the bulk density was measured by a tap density method with a GranuPack. The segregated mixtures are compared to the homogeneous ones to highlight the influence of segregation on the bulk density. A model [1] was developed to modelize the bulk density as a function of the spatial distribution of the two species in the packing. These results can be valuable in powder mixture proportioning and designing, considering the segregation impact.

References

[1] Pillitteri, S., Lumay, G., Opsomer, É. and Vandewalle, N., 2021. Forced segregation in binary granular mixtures. In EPJ Web of Conferences (Vol. 249, p. 02001). EDP Sciences.

462b Industrial Needs in Solids Handling

Tim Bell, Massih Pasha

Methods to design and troubleshoot industrial facilities handling bulk solids have seen limited practical improvement in the past 40 years. Techniques to engineer particulate products for desired attributes are still tedious and success is not guaranteed. Notable progress has been made via the advancement of automated powder flowability test equipment and mathematic modeling, but direct conversion of the results of such work to explicit solutions is rare. Many routine problems still require trial and error or empirical solutions.

This paper will review the realistic "state of the art" of solids handling technology and highlight important problems for which empirical testing is still the only, or primary, means of designing new equipment or solving problems in the field. These problems include the prediction of pneumatic conveying behavior; cohesion, wall friction and adhesion; caking; loss-in-weight feeder performance; heat transfer; grinding; and solids mixing. Comments will be provided on how these situations are addressed currently, the practical ramifications, and what may be required to make improvements.

462d Irregular Soil Particle Shape Modeling

Jennifer Curtis

NASA is interested in simulating the descent of a spacecraft approaching the Lunar and/or Martian surface along a specified trajectory which will involve gas jet-soil particle surface interaction through the landing and engine shutdown. In order to predict the erosion of the Lunar or Martian soil (regolith), a description for the solid-phase stress for the regolith is needed. For Lunar and Martian soils, the particles are highly aspherical and irregular, and can mechanically interlock, resulting in a particle bed with very high strength. These properties of the soil, and the resulting particle-particle interactions, greatly affect the rate of crater growth, the trajectories of the liberated particles, and the resulting crater shape. This presentation will outline the development of stress closures for a particle-phase continuum treatment that incorporate features of the aspherical particle shape that are typical of actual regolith.

462e Innovation Challenges in Solids Processing at Industrial Scales

Shrikant Dhodapkar

Survival and sustained growth of any business in the current environment of global market driven economies hinges on innovation and rapid adaptation. It is difficult to open a business magazine without hearing this buzzword. However, innovation is often confused with invention. While invention refers to new ideas or concepts generated through creative thought process and/or scientific research, innovation implies bringing them to market (generating value) and making a difference in people's lives. The difference is exemplified by the plethora of patents and the relative paucity of true innovations. Innovations do not always have to be breakthrough or game changing. They can be incremental and significant as long as the value generation outweighs the associated costs. Marketing pressures and the need for product differentiation often leads to namesake innovations with no tangible value. A better strategy would be to implement a few outstanding innovations rather than a smattering of mediocre ones.

This presentation, based on author's experiences and that of his peers, will highlight the challenges associated with innovation in solids processing at industrial scales. Practical aspects and best practices for being a successful innovator will be discussed.

Reflections on my career in Particle Technology

AIChE Annual Meeting, Phoenix
November 16, 2022

John W. Carson, Ph.D.
Jenike & Johanson, Inc.
jwcarson@jenike.com
978/649-3300

ABSTRACT

Using examples from 50+ years in particle technology, the author provides a brief synopsis of his career, then discusses some of the lessons he has learned followed by advice for those entering this field.

INTRODUCTION

I grew up in Houlton, a small (pop. 5,000) farming community in northern Maine – famous for potatoes. Following graduation in 1962 I entered Northeastern University (NU) in Boston.

During my freshman year at NU, I met several times with my assigned Co-op Coordinator to discuss potential job openings. During our first meeting he brought out a map of Maine to find out where Houlton is located. He told me that perhaps there would be an opening at Great Northern, a paper mill about 70 miles away. I told him that I wasn't interested in working at a paper mill and that I was willing to go anywhere there was good job. He mentioned that there were a quite a few co-op students working at US Steel's Fairless Works in Morrisville PA, and he would pursue an opening there if I was interested. I said, "Sure!"

The summer following my freshman year I went on my first co-op job to Fairless Works. Initially I worked in the engineering department of the coke plant[1], then later in Central Engineering. As a result of my work experience and classes that I took at NU, I changed my major twice: from Chemical Engineering to Industrial Engineering, then to Mechanical Engineering.

Fig. 1 Pushing coke from a coke oven battery

[1] Metallurgical coal is converted to coke by heating in enclosed ovens. This is the first step in making steel.

After two years working at Fairless Works, I knew one thing for sure: I didn't want to work in a steel mill for the rest of my life! I asked my Co-op Coordinator if there were other co-op positions available. About that time my first boss at Fairless Works was at NU interviewing graduating seniors. When he heard that I was looking to change jobs, he asked to meet with me. His first question when we met: "What do you want to do after graduation?" I told him I really didn't know for sure, but I had an interest in R&D. The next thing I knew he set up an interview for me at US Steel's Applied Research Laboratory in Monroeville PA. During my interview there I met with four groups, one of which was Bulk Solids Handling Research. At the end of day, I was asked, "If we offer you a co-op job here, which group do you want to work in?" I replied, "I'll work in any of them – they all sound interesting."

Fig. 2 US Steel's Applied Research Laboratory

They offered me a position, and I accepted. I didn't know which group I would be working in until I showed up for my first day of work and was told, "We've decided to place you in the Bulk Solids Handling Research group." This was a lively, interesting group of individuals led by Henk Colijn, with Jerry Johanson as Chief Researcher. I did research on loads acting on inserts inside bins, studying if differences in wall friction between fine and coarse particles could be used to separate particles exiting from a sloping plate, and other topics.

Johanson left US Steel in 1966 to form Jenike & Johanson, Inc. ("J&J"), while I stayed to the end of my co-op in December 1966 and graduated from Northeastern in June 1967.

Fig. 3 Conducting my first Jenike Shear Test

Since J&J's office was in Winchester MA (a suburb of Boston), I had the opportunity to meet Andy Jenike in the spring of 1967, and he offered me a summer job between undergrad and grad school. The company's office and lab were located in the basement of Jenike's house.

I then went on to MIT where I received my MS in Mechanical Engineering in September 1968 and my PhD in Mechanical Engineering in October 1970.

Upon graduation from NU, I had been commissioned in the US Army through the ROTC program. I applied yearly for a delay to active duty so that I could attend grad school. Each year's request was granted until June 1970 when I was told, "No more delays after this year". I hurriedly completed my PhD in October of that year.

Expecting that I had at most six to eight months before I'd have to go into the Army, I didn't interview any companies but instead took a "short term" job with J&J upon graduation. In the spring of 1971, I received a letter from Uncle Sam saying in effect, "We have too many officers coming into the Army because the Vietnam War is winding down. Would you be interested in serving 3 to 6 months Active Duty for Training followed by a 2-year extension of your Army Reserve commitment?" I went to Jenike and asked if it was OK for me to stay. He turned my question back to me: "Do you want to stay?" I said, "Yes" -- and 52+ years later I'm still there!

Fig. 4 Capt. John Carson

I served as President of J&J from 1985 until I retired from full-time employment in 2018. The firm grew from two rooms in the basement of Jenike's house in Winchester MA when I started to seven offices around the world – three in the US plus one each in Australia, Brazil, Canada, and Chile. We have the enviable reputation of being the largest and most experienced consulting engineering firm in the field of storage, flow, and processing of powders and bulk solids in the world.

SOME LESSONS LEARNED

Lesson #1: Bulk solids handling is still a relatively unknown field to most engineers

Bulk solids handling is, by far, the world's largest industrial activity and certainly one of the most mature, having been carried out for over 2,000 years. It has been estimated that over 16 billion tons of common bulk solids are handled – often many times – every year. [1] Yet, even with all this history and wealth of experience, far too many solids handling projects still go wrong!

A strong reason for the lack of understanding about this technology is that education of engineers in this field isn't much better than it was 40 or 50 years ago. A tiny fraction of the engineers graduating with a Bachelor of Science degree anywhere in the world receive even some introductory training in bulk solids handling.

We at J&J have been offering short courses in this technology through the American Institute of Chemical Engineers (AIChE) and other groups since the early 1970s. Despite the popularity of these courses, only a small fraction of engineers in industry receive such training. Most of the rest often flounder when faced with a bulk solids handling problem. This is even truer in developing areas of world, so they are ripe for introduction of this technology.

In the 1980s we started mass producing a small, plastic mass flow - funnel flow model. To date we have given away over 50,000 of them. I'm thrilled when someone holds with the model for the first time and, in amazement, learns that simply steepening the hopper walls changes the flow pattern from funnel flow to mass flow and thereby eliminates ratholing, increases live capacity, lessens the likelihood of arching, and minimizes particle segregation.

Another example of the lack of understanding of bulk solids handling principles involves material-induced loads on bins and silos. Not only are most of the codes woefully inadequate, but mistakes in design and operation are commonplace. For example, additional discharge points are often added to bins and silos without any thought being given to how they change the flow pattern and material-induced loads. It's as if the engineer thinks that the structure is holding a liquid, in which case adding another outlet has no such effect.

Lesson #2: Merrow's findings are still valid

In the early 1980's, the US Department of Energy commissioned the RAND Corporation to investigate the performance of 37 US and Canadian-based processing plants with a view to understanding performance. Edward Merrow led a group that studied plants involved chemical transformations to some degree, and some featured new chemical routes; there were no new unit operations involved. The value chains studied also covered the full spectrum of feed, intermediate and product states (liquids, gases and solids).

The high level findings from this very insightful study [2] include:

- Value chains featuring bulk solids as the mainstream performed worse than plants featuring liquids or gases only.
- Two-thirds of these operations (featuring bulk solids) operated at less than 80% of their nameplate (design) capacity at the end of their first year, with a quarter failing to reach even 40%. The average for this cohort was 64% of nameplate compared to 90-95% for the liquid and gas operations.
- In terms of the root causes of these failures, lack of knowledge around heat and mass transfer was the biggest factor. While there are many heroes of fluid flow, the field of bulk solids flow, which is arguably as important and ubiquitous, is less well researched. As a consequence, our understanding of the transport phenomenon associated with bulk solids is poor when compared to gases and liquids.
- Of the 37 plants studied, only 6% had no major performance issues (defined as a shut-down for one or more weeks during the first year), and the overwhelming majority of problems involved related to process mechanics (bulk solids behavior) as opposed to chemistry (as per the Ravensthorpe Nickel Operation case study below).
- When compared to plants built in the 1960's (20 years prior) there was no difference (i.e., signs of improvement). The improvement effort during that period was found to be focused on process chemistry as opposed to mechanical processes (understanding flow and doing the bulk solids science homework).

These were considered surprising and disturbing findings at the time, and in response Merrow recommended:

- More basic research in support of understanding the behavior of bulk solids especially in a process environment.
- Characterization of the actual bulk solids involved in the proposed process, warning against using "library values" that fail to capture the nuances that control the behavior of bulk solids.
- Feedback (operations back to designers) and clearer accountability for bulk solids flow and behavior issues in designs. The myth that any physical flow problems can be simply dealt with by the operation on start-up needs to be killed.
- Need to give the behavior of bulk solids equal billing with chemistry in terms of research and development time, debunking the myth that physical flow is trite and not worthy of attention.

Move forward another 20 years and in 2005 Bell [3] raised the question,

> *Given the low plant productivity and long startup times associated with not taking Merrow's advice, why would anyone build a new plant without fully piloting the design*?

He noted that this question is at the center of many capital project decisions and figures prominently in the post-mortem of failed projects. He then listed three probable reasons for not piloting a process:

1. Ignorance of the work of Merrow and others.
2. Pride. He observed that Merrow's data represents averages, and most engineers consider themselves above average. Many of them believe they are clever enough to beat the odds.
3. A legitimate need for innovation and a haste to get new products or processes to the marketplace. The construction, operation, and refinement of a pilot plant will take many months and can cost millions of dollars. Business management may conclude that a business opportunity window will no longer be open if the process development timeline is too long.

Bell's work is a staggering finding that shows that the dial had not moved at all as a result of Merrow's [2] and [4] findings and recommendations. This finding was made even more poignant by the fact it represented an even wider range of industries (food, nutraceutical, pharmaceutical, chemical, building materials, minerals processing, etc.). The common denominator in failure remained bulk solids. The common root cause of this was designers failing to do their homework and differentiate between the flow of liquids and bulk solids.

I've personally seen this repeated time after time during my career. I recall when Circulating Fluidized Bed technology was being commercialized. This technique for steam generation allows virtually any material with calorific value to be used. Many of the initial applications involved siting the plants near piles (mountains, actually) of coal refuse (bituminous gob, anthracite culm). These materials worked well once they were inside the fluidized bed but getting them in there was a major challenge that, unfortunately, had been overlooked. As a result, many of these plants took far longer to reach rated capacity and suffered significant cost overruns.

These same problems are being repeated more recently in projects involving biofuels, additive manufacturing, and battery manufacturing to name but a few.

To help make these value impacts more meaningful, it is useful to consider a case study. The Ravensthorpe Nickle Operation (RNO) project is a very relevant, relatively recent and poignant case study in lack of understanding of bulk solids handling technology.

As described by Wellwood [5], RNO was a flagship project for the world's largest mining house, BHP Billiton (BHPB). The Basis of Design (BoD) was to produce 50,000tpa Ni equivalents from a Saprolite (30%)/Limonite (70%) ore via a new combined (pressurized acid leach and atmospheric leach) leaching step called Enhanced Pressure Acid Leach. The scope of the project included development of a mine, processing plant and associated infrastructure near Ravensthorpe in Western Australia. The goal was to produce mixed nickel cobalt hydroxide intermediate product for processing at the company's existing Yabulu refinery, whose accommodating capacity expansion was also in the scope of this mega-project.

In 1997, the project entered a study phase which was to last seven years and pass through the standard pre-feasibility and feasibility stages. The salient features/achievements of the completed RNO study phase included:

- $85 Million Front End Loading (FEL)
- 200,000 engineering-hours
- Eight months continuous pilot plant operation
- Evaluation of ore blends covering Life of Mine (20y)
- Five independent technical reviews
- Rigorous internal processes (including hydrometallurgical expertise from alumina business)
- Detailed mine planning

RNO emerged from its final study phase gate (4) in March 2004, with approval to proceed at the cost of US$1.04 billion. In the context of this announcement, the project was promoted to investors as having the best possible profile for any greenfield laterite project as it was expected to provide the high leach feed grade with low mining costs.

The detailed design and construction phase of the project started immediately thereafter with the appointment of two, world class Engineering-Procurement-Construction-Management (EPCMs). Unfortunately for everyone involved -- directly and indirectly -- in 2009, BHP Billiton announced that it was suspending production at the Ravensthorpe nickel mine indefinitely as this mega-project had failed on all its metrics (Table 1 below).

At the point of its suspension in 2009, RNO has only reached 20% of its design capacity and was registered as a $3.615 billion impairment to BHPB with over 1,800 jobs lost. The Ravensthorpe operation was eventually sold to First Quantum Minerals Australia Pty Ltd, for US$340 million and, after an extensive 18 months of modifications and refurbishment, was operated profitably.

Table 1. Summary of RNO metrics		
Performance metric	**Expected**	**Actual (at point of suspension in 2009)**
Capital cost	$1.04 billion	$2.1 billion (100% overrun)
First production	2006	2008 (18 months late)
Nameplate	50,000 t Ni eq/yr	20% nameplate

While there were undoubtedly market-related issues [6], the root cause of the failure was really a technical one: the operation's inability to achieve its nameplate throughput in a sustainable manner. Although the core transformational step in the operation (combined acid and pressurized leach) performed exceptionally well, it was starved of feed (maximum sustainable throughput only 20% of BoD nameplate capacity) because of front end materials handling problems baked into the final design. [7, 8]

So how did a project managed by the world's largest mining house and with such an extensive study phase fail so spectacularly, from what was subsequently revealed as known and demonstrably addressable causes?

Process engineering is in essence a dynamic field where flow (particulate solids, liquids and/or gases) through the value chain underpins everything that most engineers do. However, for most engineers, the fact that bulk solids flow differently from liquids is a classic "blind spot" akin to that found in psychology associated with a Johari window.

The Johari window is a technique [9] designed to help people better understand their relationship with themselves and others. Created by psychologists Joseph Luft and Harrington Ingham in 1955, it is used primarily in self-help groups and corporate settings as a heuristic exercise. Luft and Ingham named their model "Johari" using a combination of their first names.

In the exercise, someone picks a number of adjectives from a list, choosing ones they feel describe their own personality. The subject's peers then get the same list, and each picks an equal number of adjectives that describe the subject. These adjectives are then inserted into a two-by-two grid of four cells. See Fig. 5.

The philosopher Charles Handy called this concept the "Johari House with four rooms". Room one is the part of ourselves that we and others see. Room two (the "blind spot") contains aspects that others see but we are unaware of. Room three is the private space we know but hide from others. Room four is the unconscious part of us that neither ourselves nor others see.

A relatively small but diverse group of engineers around the world knows the technology of bulk solids handling and how to apply it. This is the "known by others" group. The vast majority of engineers, unfortunately, is unaware of even the existence of this technology, so they fall into the "unknown by self" group. This combination results in the "blind spot" that helps to explain classic failures such as RNO.

	Known by Self	Unknown by Self
Known by others	open/free area	blind spot
Unknown by others	hidden area	unknown

Fig. 5. Johari window

ADVICE FOR THOSE ENTERING THIS FIELD

Advice #1: Find a mentor – hopefully, you'll learn from their experiences

I was very fortunate to work closely with Andy Jenike for over a decade. He set high standards for technical accuracy and ethics, and he was fair, honest, and a very hard worker. [10]

In addition to being a first-class engineer and mathematician, Jenike had an excellent business sense. Perhaps one reflection of this is that the company that he and Jerry Johanson founded is still going strong for over 55 years. This is in no small part the result of Jenike's keen sense of what clients need and how best to provide that service. He insisted on quality in everything he did, and this rubbed off on everyone around him.

Jenike had high ethical standards. He recognized that in order to maintain the confidence of clients it was essential that the advice offered be completely unbiased. I recall an instance when we had recommended that a client go to a certain equipment supplier for some custom-built equipment. The supplier was so appreciative that he sent us a check as a finder's fee. Even though our client was unaware of this payment, Jenike returned it because he didn't want even the perception that his recommendations might be biased.

Finding and retaining quality employees is important for any business, but Jenike's approach as president of a small company was unique. He recognized that we were competing against major firms in the field, not just small, local engineering companies, so, in order to attract and retain employees, he insisted that we provide benefits comparable to large firms. Furthermore, he wanted those benefits to provide protection to employees and their families to the full extent allowed by the IRS, so employees didn't have to worry about paying their bills if something unexpected happened. We have retained these policies even today.

As a result of policies such as these, J&J has an enviable record of employee retention. Many have been with the company all of their career, while many others who have not yet retired have in excess of 30 years with the company.

Although Jenike was generous with his employees, he was a frugal man. He would use an eraser until it was worn down such that he could barely hold it between his fingers.

Jenike was a very down-to-earth man. Despite of his imposing stature and European charm, he insisted on being called "Andy". No job was beneath him. As is typical in a small company, our employees were asked to wear many hats, and Jenike did his share as well. I recall him going out with a pair of scissors to cut a small grass strip in front of our office before a client showed up. His attitude of not putting himself above any job that needed to get done helped foster the idea that everyone had a stake in the company and had an equal share in its success. He insisted on limiting stock ownership to employees, with no outside investors.

Jenike was a pioneer in not only technical matters but also in business. We established a no-smoking policy in the early 1970s, well before it became fashionable. Jenike had wanted to do this for some time, but a few employees were smokers and found it hard to quit. Changing the policy became easier to implement when he found one of our lab technicians with a lit cigarette while running a shear test in our high temperature oven with hydrogen gas being fed in.

Advice #2: Collaborate with others who have similar interests and drive – this is a team sport!!

Given the complexity and non-intuitive nature of bulk solids handling problems, teamwork and collaboration are essential.

The success of a company like J&J – and indeed most companies and organizations -- is the result of many people, not just the leader. In the mind of some CEOs – and they're usually not shy in making this known to anyone who is within earshot – every original idea and technical insight originates with them. I knew from working with Jenike, and I knew even more strongly after serving nearly 35 years as President, that the success of J&J is the result of the collective effort of all our employees.

It is interesting to note that, of the numerous academic and industrial groups around the world who have made major contributions to our field over the past 60 years, only two remain strong today: J&J and TUNRA[2] in Australia. I attribute this to the leadership of both organizations: Jenike and those of us who followed him, and Prof. Alan Roberts at TUNRA. As brilliant and accomplished as Alan is, he has always been quick to recognize the contributions of others at TUNRA. I made this

---

[2] The University of Newcastle Research Associates

observation to him a number of years ago while on a taxi ride in Edinburgh along with Prof. Michael Rotter. Alan, in his typical fashion, smiled and nodded his head.

There is a great group of academics and industrial folks who have passion for this field, and I have been fortunate to know and work with many who have built this field to where it is today. This includes a number who have passed away: Henk Colijn, Reg Davies, Abraham Goldberg, Pedrag Marjanovic, Jan Novosad, Ivan Peschl, Fritz Rademacher, Brian Scarlet, Jörg Schwedes, Sunil de Silva, Fred Thomson, Jürgen Tomas, Ugur Tuzun, John Williams, Harold Wright.

Another important group who I have known and worked with includes those who have retired or semi-retired: Peter Arnold, Gisle Enstad, David Goodwill, Richard Jenkyn, Jerry Johanson, Alan Roberts, Michael Rotter.

Groups such as AIChE's Particle Technology Forum, ASTM's Subcommittee D18.24 Characterization and Handling of Powders and Bulk Solids, and the European Federation of Chemical Engineering's Working Party on Mechanics of Particulate Solids provide excellent venues by which to collaborate with others and advance the field of knowledge. There are also many industry-focused organizations (e.g., American Association of Pharmaceutical Scientists, International Society for Pharmaceutical Engineers, Metal Powder Industries Federation). that include bulk solids technology as an area of focus. The fact that few of these organizations have many members with knowledge of bulk solids technology means that there is an opportunity for those with expertise (even "some") in this field to lead and/or connect in a collaborative way.

Advice #3: Get "real world" (industrial) experience

Bulk solids behavior is complex, so the movement of particles in full-scale industrial situations is essential to keep oneself grounded.

Jenike was firmly of the opinion that experiments are best used to verify or modify a theory, and that experiments without a theory have very limited use. One of his favorite expressions was, "Whenever you have an urge to run an experiment, lay down and take a nap. Usually, the urge will go away." What he meant was that one's time is better spent coming up with a comprehensive theoretical model.

Those whose knowledge of this field is limited to theoretical or computational models usually don't know the limits of their models, where to apply -- or not apply -- them, or what "real world" behavior looks like.

On the other hand, there are many individuals who have "real world" experience (engineers and operators) but have no idea or knowledge as to "why" they are seeing what is happening.

The world needs people with a diverse/wide set of experiences and knowledge to successfully solve and prevent bulk solids problems, never mind to truly add to this field.

Advice #4: Don't be afraid to take chances, and don't let mistakes and failures discourage or define you – learn from them instead

You may not be successful, but you won't know unless you try.

Mistakes and failure are part of life. This is particularly true in one's personal life. Serious illness, death of a family member or close friend, loss of a job, financial distress affect almost everyone at one or more times during our life. The same is usually true in one's professional life. If you don't experience them, you probably aren't trying hard enough!

It is important to assess risk when trying something new, to be sure you understand who may be impacted if things don't go as planned, and to have thought this through. In other words, plan ahead on how to detect and react if your assumptions are not right.

The common reaction to mistakes and failures is to become defensive, blame others, or try to hide in a hole. That might feel good for a time, but it solves nothing. The important thing is to learn from your experiences, so you don't repeat them in the future.

Advice #5: Continue to push boundaries of knowledge

The authors of a 2005 NASA paper [11] observed:

> *Working with soil, sand, powders, ores, cement, etc. and using hoppers are so routine, that it seems straightforward to do it on the Moon and Mars as we do it on Earth. This paper brings to the fore how little these processes are understood and the millennia-long trial-and-error practices that lead to today's massive over-design, high failure rate, and extensive incremental scaling up of industrial processes because of the inadequate predictive tools for design.*

My friend Lyn Bates of Ajax Equipment has been quoted as saying:

> *Bulk solids technology is not rocket science – it's much more difficult than that! What's more, rocket scientists agree!*

The number of problems to be solved is endless. Some of the many common ones include:

- Ratholing
- Caking
- Segregation
- Silo vibrations
- Properties of & design procedures for unusual bulk solids (e.g., anisotropic, viscous & cohesive, both elastic & cohesive)

Advice #6: Get involved with developers of silo codes

Bins and silos fail with a frequency greater than any other structure. [12] A failing bin can release huge potential energy that can kill or seriously injure anyone nearby as well as cause significant property damage and plant interruption.

Part of the reason for this sorry state of affairs is that most Codes and Standards in this field are very weak, especially in No. America. For example, the American Concrete Institute's "Design specification for concrete silos and stacking tubes for storing granular materials (ACI 313-16) and Commentary" [13] includes a table of typical flow properties. A common bulk solid, bituminous coal, is listed as having a coefficient of friction against steel of 0.30. The reader is provided no guidance as to the type of bituminous coal, its moisture, ash, or other characteristics for which this value applies. Furthermore, there is no indication as to the type of steel: smooth carbon steel, rusty carbon steel, stainless steel (what type, what surface finish?). The reader is left with the impression that none of these variables is important – which is patently false. The footnote warns in small print,

> *Design parameters should preferably be determined by test and the values above used with caution.*

I suspect that most users of this Standard overlook this warning.

The situation is similar with ANSI/ASAE S433.1 JAN2019 "Loads exerted by free-flowing grain on bins" [14] and the National Farm Building Code of Canada 1995. [15] While both are quite limited in scope, the guidance provided is somewhat simplistic and easily misused.

Over 20 years ago the American Society of Mechanical Engineers formed a committee to develop a standard for bin and silo design. Called "Structures for Bulk Solids", this document is getting close but has not yet received final approval. However, if/when it does it will be likely be extremely limited in the conditions that it covers. For example, the current draft limits the container shapes to:

> *...a cylindrical section that is circular in cross-section, and a converging hopper portion beneath the cylinder that is either a plane flow wedge (with vertical end walls) or a right circular cone. The geometry is further defined to be symmetrical.*

Bins and silos with multiple and/or off-centered outlets, transition hoppers, and other common silo configurations will not be covered by this standard.

The most comprehensive code currently in existence is British Standard BS EN 1994-1:2006 "Eurocode 1 --Actions on structures – Part 4: Silos and tanks". [16] Yet, as complete as this code is, it still does not cover many common design conditions faced by users. [17]

Given the simplistic and often incorrect guidance provided by these and other codes coupled with the fact that bins and silos continue to fail, it is imperative that those who have knowledge of bulk solids handling principles get involved – and remain involved despite the delays and frustrations inherent in this work.

Advice #7: DEM is a great tool, but use with caution

Discrete Element Modeling (DEM) has come a long way since the turn of the century. Twenty years ago, the state-of-the-art was limited to spherical particles of varying size. Fifteen years ago, "gluing" spheres together to create clusters was introduced. Today, modeling of more realistic particle shapes using polyhedra is common.

While this technology holds huge potential, it is still a long way from modeling most "real" systems. For example, the number of particles in an industrial-size silo is on the order of 10^{15}, but current DEM codes are limited to on the order of 10^{6} particles for realistic run times – nine orders of magnitude difference! At present the demands on computing power are impractical, particularly considering that the time scale is inversely proportional to particle size. Then there is the very real problem of particle cohesiveness, which has yet to be fully addressed. Furthermore, as particles get smaller (real world or model), the effects of surrounding air or other gas become significant, so they must be included.

One must guard against assuming that pretty pictures and videos that "look right" are sufficient to ensure that the results of a DEM run are correct. For example, consider the two images shown in Fig. 6. Both involve simulation of flow from a 45° conical hopper. The left figure was generated using spherical particles with low friction, while the one on the right was generated using polyhedral particles with more realistic friction. Anyone who has experience with hopper flow will quickly recognize that the simulation on the right is much closer to most real world hopper flows than the one on the left.

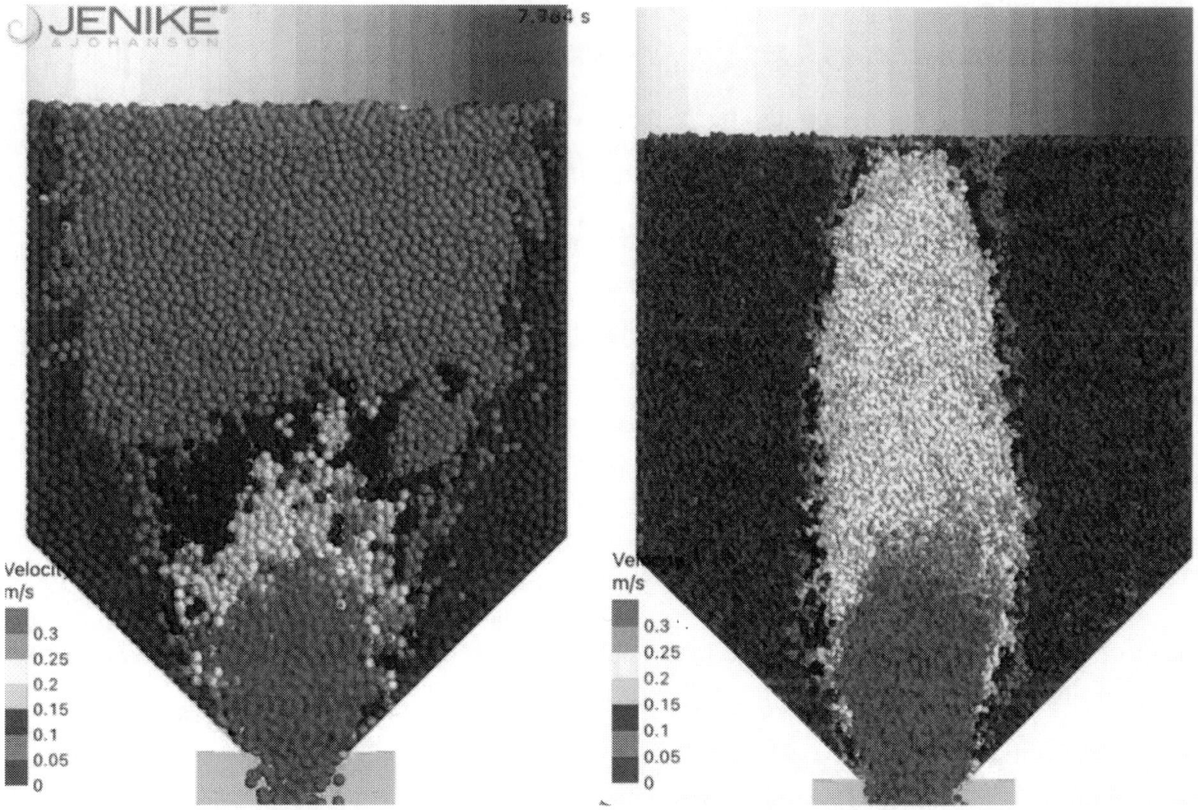

Fig. 6. Left: generated using spherical particles with low friction. Right: generated using polyhedral particles with more realistic friction

It must be remembered that, at best, DEM can answer the question, "Why is this problem occurring"? It cannot provide guidance as to what to change to correct the problem. This is where the expertise of design engineers who have worked in this field for years comes into play.

Advice #8: Work to make a difference in the world

More than many other fields today, bulk solids handling technology offers the opportunity for one to make a significant impact on the world around us – both near and far.

One area where bulk solids technology can make a difference involves the development of new materials, such as:

- fertilizers that improve crop yield while doing less harm to the environment
- foods that are more healthy and less expensive

- pharmaceuticals (tablets and capsules) that cure or prevent diseases
- means for energy generation that are more efficient and less harmful to the environment
- batteries that provide more power at lower cost
- means to effectively and efficiently capture carbon from the atmosphere

Another area where bulk solids technology can make a difference involves improved plant operation, such as:

- Safety so workers don't have to swing sledgehammers to get material to discharge, and they don't run the risk of serious injury or death from a collapsing bin or dust explosion
- Efficiency so plants are more profitable
- Waste reduction so there's less harm to the environment
- Quality so there are fewer rejects and customer complaints
- Productivity so production rates are higher

Working to make a difference in the world is one of the keys to having a meaningful life. In this regard, my philosophy of life is neatly summed up by the following (source unknown):

The day you were born, you cried -- but the world rejoiced.

Live your life so that the day you die, you will rejoice -- but the world will cry.

SUMMARY AND CONCLUSIONS

Looking forward I am confident that the future is bright for those individuals who are currently working in this field and for those who will join in the future.

Bulk solids handling is still a relatively unknown field to most engineers, so plants are still being designed and built that suffer from lack of this knowledge.

For those just starting out in this field, I offer the following advice:

- Find a mentor – hopefully, you'll learn from their experiences
- Collaborate with others who have similar interests and drive – this is a team sport!!
- Get "real world" (industrial) experience
- Don't be afraid to take chances, and don't let mistakes and failures discourage or define you – learn from them instead
- Continue to push boundaries of knowledge since there are many unsolved problems needing attention
- Get involved with developers of silo codes
- DEM is a great tool, but use with caution
- Work to make a difference in the world

REFERENCES

1. Taylor, T., "The global status of bulk materials handling 2013", British Material Handling Board, edited by Lyn Bates
2. Merrow, E., "A quantitative assessment of R&D requirements for solids processing technologies", Available at: https://www.rand.org/content/dam/rand/pubs/reports/2006/R3216.pdf (Accessed: 6 March 2019), 1986
3. Bell, T. A., "Challenges in the scale-up of particulate process – An industrial perspective", Powder Technology, 60-7, 2005
4. Merrow, E., "Problems and progress in particle processing", Chemical Innovation, Jan. 2000 and Chemical Engineering; Oct. 1988, Vol. 95, Issue 15
5. Wellwood, G. A., "Bulk solids handling: Perspective on a professional blind spot", The Chemical Engineer, Flow and Tell issue, January 2020, pp. 24-31
6. Department of State Development, "Submission to Parliamentary Accounts Committee: Ravensthorpe Nickel Operation", Perth: Government of Western Australia, 2010
7. Piper, D., "Ravensthorpe: From white elephant to resurrection", Australia's Paydirt, Vol. 1, No. 193, 16-23, April 2012
8. Moore, P., "Nickel future for FQM". International Mining, pp. 70-75, Oct. 2011
9. Luft, J. and Ingham, H., "The Johari window, a graphic model of interpersonal awareness". Proceedings of the Western Training Laboratory in Group Development, Los Angeles: *University of California, Los Angeles*, 1955
10. Carson, J.W., "Andrew Jenike, A true visionary in particle technology", 7th World Congress on Particle Technology (WCPT7), Procedia Engineering 102 (2015) 657 – 660
11. R. A. Wilkinson, R.A., Behringer, R.P., Jenkins. J.T., and Louge, M.Y., "Granular materials and the risks they pose for success on the Moon and Mars", in Proceedings, STAIF, AIP Conference, vol. 746 (2005), pp. 1216-1223
12. Carson, J.W., "Silo failures: Case histories and lessons learned", Proceedings of 3rd Israeli Conference for Conveying and Handling of Particulate Solids, Vol. 1, p. 4.1-4.11, 2000
13. ACI 313-16 "Design specification for concrete silos and stacking tubes for storing granular materials (ACI 313-16) and Commentary", American Concrete Institute, 2016
14. ANSI/ASAE S433.1 JAN2019 "Loads exerted by free-flowing grain on bins", Jan 2019
15. National Farm Building Code of Canada 1995, Issued by Canadian Commission on Building and Fire Codes, National Research Council of Canada, 1995
16. BS EN 1991-4:2006 "Eurocode 1 – Actions on structures – Part 4: Silos and Tanks", British Standard, 2006
17. Carson, J.W. and Craig, D.A., "Silo design codes: Their limits and inconsistencies", presented at 7th World Congress on Particle Technology (WCPT7), May 19-22, 2014, Beijing, China, Procedia Engineering, Published by Elsevier Ltd. (open access article under the CC BY-NC-ND license)

408a Development of an Online Free Web Application to Perform Multivariate Analysis and Determine Microcrystalline Cellulose Crystallinity from Raman Spectra

Ana L. P. Queiroz, Abina Crean

Introduction

Strategies to understand and mitigate batch-to-batch raw material variability are key to ensuring process and product compliance. When critical material attributes are well-understood and quantified, operating strategies can be devised to ensure that the process will be able to cope with incoming material variability. Microcrystalline cellulose (MCC) is a raw material widely used across different industry sectors, including food and pharmaceutical. This is a semi-crystalline substance with inherent variable crystallinity (%CI) due to differences in raw material source and the different manufacturing processes through which MCC can be produced. This variance in MCC crystallinity can result in downstream process variability and should be well understood and controlled.

Aim

The aim of this study was to develop models to determine the MCC crystallinity index (%CI) from Raman spectra of 30 commercial batches using Raman probes with spot sizes of 100 mm (Kaiser Raman MR probe) and 6 mm (Kaiser Raman PhAT probe). The second aim was to pack the data and the designed models in a user-friendly online free-access tool (McCrystal – crystallinity research©) to disseminate the models to users with diverse levels of modeling experience.

Experimental

30 commercial batches of MCC were ball milled to produce amorphous samples – confirmed by Powder X-Ray Diffraction. Then, pellets of the batches as received and of the corresponding amorphous material were prepared. The pressure used to prepare these pellets was just enough to hold the powder pile together without compacting it.

Spectra of all pellets were acquired using Raman probes with spot sizes of 100 mm (Kaiser Raman MR probe, sampled on 3 different areas on one surface of the pellet) and 6 mm (Kaiser Raman PhAT probe, sampled on 1 area on the top and 1 area on the bottom of the pellet). Spectra fluorescence and baseline shifts were removed by linear-interpolation baseline and standard normal variate (SNV) normalization.

The %CI was determined using a univariate model based on the ratio of the Raman peaks at 380 and 1096 cm^{-1}. The univariate model was adjusted for each probe. The %CI was also predicted from spectral data from each probe using partial least squares regression (PLS) models (where Raman spectra and univariate %CI were the dependent and independent variables, respectively). PLS model was initially performed using The Unscrambler X. However, all models were re-written in R, and the code written in R was compiled with the data used to build the models to be shared with the scientific community.

Results

Both univariate and PLS models showed good predictive power; the univariate model had a residual sum of squares (RSS) of 0.052 and 0.007 and Pearson's r of 0.895 and 0.969 for the MR and the PhAT probes, respectively, and the PLS models had a validation explained variance of 97.47 and 97.16% Bias of 0.015 and - 0.068 for the MR and the PhAT probes, respectively.

The developed PLS model substantially reduced the analysis time as it eliminates the need for spectral deconvolution.

A principal component analysis (PCA) model was used to separate the Raman spectra of the same samples captured using the different probes. This can be useful to identify non-conformity across instruments, which were encountered in this work, but also across different batches.

Conclusions

Crystallinity index was determined for 30 commercial batches of MCC using two different models, a univariate model of the ratio between two peaks and a PLS regression. A user-friendly online free-access tool, namely McCrystal–crystallinity research©, was designed to determine crystallinity of new batches based on the models proposed in this study and Raman spectral data. McCrystal also performs MCC spectral baseline by linear interpolation, spectral normalization by SNV, and PCA. The PCA model included in this web application enables fast identification of non-conforming batches of MCC based on batch-to-batch crystallinity variation.

The web application is available at: https://sspc.ie/mccrystal-registration/

408b Influence of the Feed Frame Design on a Continuous Manufacturing Process

Peter Boehling, Johan Remmelgas, Julia Kruisz, Johannes Poms, Manel Bautista, Johannes G. Khinast, Emmanuela Gavi

The pharmaceutical industry is moving towards continuous manufacturing to replace traditional batch production. One example of a continuous direct compaction process is powder feeders that dispense materials into a blender upstream of a tablet press. Although the feeders dispense materials accurately and precisely, small variations in the raw materials or the performance of the feeders can cause small variations in the composition of the powder blend.

The purpose of the blender is to provide a homogeneous material mixture, but it is well known that considerable mixing can occur also in the tablet press. In order to understand the degree of mixing that occurs in the tablet press, DEM simulations of different feed frames are carried out. The focus is on the material flow in the feed frame and into the dies, not the compression step itself. The DEM simulations use a DEM model with calibrated material parameters of a real powder blend; this calibration routine is described in [1,2], where it has been successfully validated for a continuous mixing process.

DEM simulations are carried out for a range of feed frame designs. In addition to providing the residence time distribution (RTD), the results from the DEM simulations show how the material flows in the feed frame and provide an understanding of how each feed frame should be modeled in terms of reduced-order models that accurately capture the dynamics when the holdup mass can vary.

The reduced-order models for the feed frames are then used to show how the mixing characteristics of typical mixers can be used to specify requirements on the feeders. Or alternatively, it is shown how the feeding performance of typical feeders affect the required performance of the mixer. Since the reduced order model properly accounts for a varying holdup mass, particular emphasis is put on identifying combinations of mixers and feed frames that not only are able to handle feeder fluctuations but that also minimized material losses during startup and shutdown.

REFERENCES

[1] Toson et al., 2018. 'Detailed modeling and process design of an advanced continuous powder mixer'. Int J Pharm 552, 288–300. https://doi.org/10.1016/j.ijpharm.2018.09.032

[2] Toson et al. 'Continuous Powder Mixing Technology: Validation of the DEM Model'. Submitted to J Pharm Sci.

408c Using Multi-Level Coarse-Grained Simulation for Design Studies: The Case of Tablet Press Feeder

Lokeshwar Mahto, Tarun De, Jayanta Chakraborty, Jitendra Kumar, Anurag Tripathi, William Ketterhagen, Maitraye Sen, Krishnendu Basak

DEM simulations often produce only qualitative predictions. One of the reasons for this limitation is the inability to use particles of appropriate size. Coarse-grained particles are used in most cases for DEM simulations to save computation time. However, a recently published method can enable us to use resolved (actual size) particles in crucial regions of the machinery and thereby obtain quantitative predictions [1].

In this presentation, we shall apply multi-level coarse-graining for simulating high speed tableting machines which are frequently used in the pharmaceutical industry. The critical quality attributes of tablets depend on various design parameters such as paddle wheel shape and feed frame design and the effect of these parameters on performance of this machine will be studied using both multi-level coarse-graining and fixed coarse-graining technique and these two will be compared.

It is observed that major change in the prediction occurs when very small particles are used in the critical region using adaptive coarse graining. However, this change does not alter the design decisions since the trend predicted by the uniform coarse-grained simulation is preserved in the multi-level coarse-grained simulations [2].

Figure 1A. shows the results for the prediction of the mean tablet mass for two different paddle wheel designs. It can be seen that like the case of fixed coarse-grained simulations, the mean tablet mass does not change significantly upon change of paddle wheel. However, the absolute value is 15% higher for both cases. Similar trend can be seen for the tablet mass variation as shown in Figure 1B. Comparisons with resolved simulations will also be shown for a few cases as shown in Figure 1.

1. T. De, J. Chakraborty, J. Kumar, A. Tripathi, M. Sen, W. R. Ketterhagen, "A particle location based multi-level coarse-graining technique for Discrete Element Method (DEM) simulation', Powder Technology, vol. 398, 117058, 2020.
2. W. R. Ketterhagen, "Simulation of powder flow in a lab-scale tablet press feed frame: effects of design and operating parameters on measures of tablet quality," Powder Technology, vol. 275, pp. 361–374, 2015.

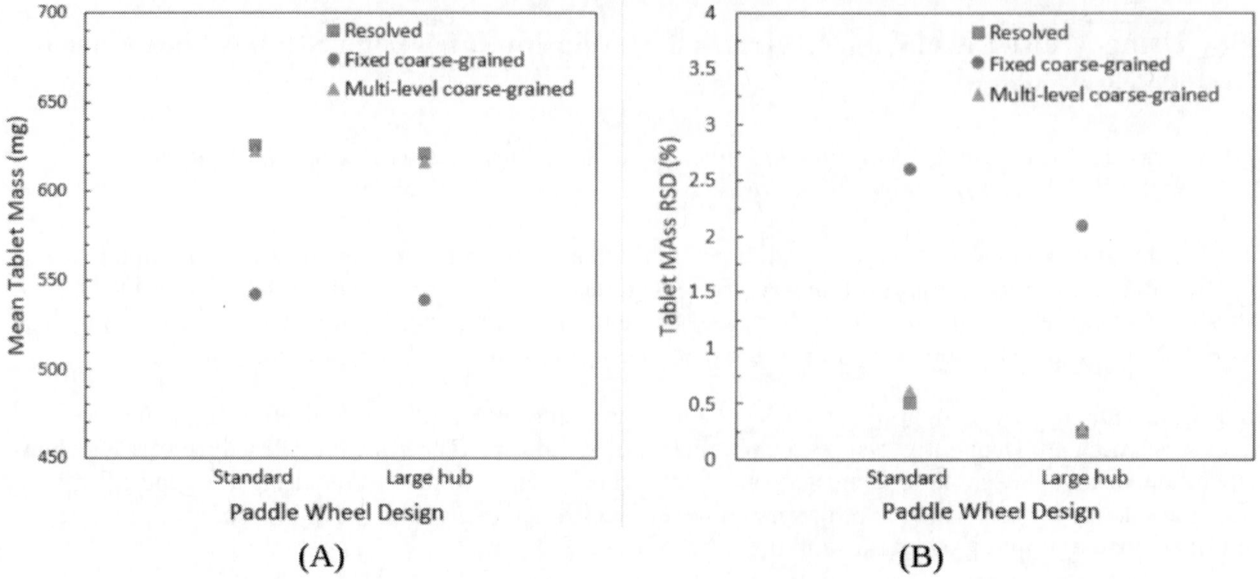

Fig.1. Tablet mass (A) mean values (B) relative standard deviation (RSD) for the given paddle shape and coarse-grained simulation

408d First Steps Toward Large-Scale Production of PLGA Nanoparticles

Carolin Tetyczka, Ramona Jeitler, Lukas Vergeiner, Daniela Fiedler, Dagmar Kolb, Gerd Leitinger, Eva Roblegg

Owing to its excellent biocompatibility, adjustable degradation and release characteristics, low immunogenicity, and high versatility, poly(lactic-co-glycolic acid) (PLGA) is one of the most widely employed polymers in the biomedical field. Consequently, more than 60 PLGA-based formulations with different properties have received US Food and Drug Administration (FDA) and European Medicines Agency (EMA) approvals, respectively. However, despite intensive research, no nanopharmaceutical based on PLGA has been granted a marketing authorization yet [1]. This can be attributed to the fact that a nanoproduct launch involves a complex and expensive pathway from formulation development on a laboratory-scale to scale-up manufacturing and clinical trials. Thereby, scale-up is particularly challenging, since nanoparticle characteristics including their efficacy and safety, are directly related to their physicochemical properties and may change during conversion from laboratory-scale to large-scale production [2], [3]. In addition, laboratory-scale manufacturing of PLGA nanoparticles is mainly conducted by bottom-up methods such as nanoprecipitation, dialysis and ultrasonication, which involve a variety of different processing steps and are not transferable to industry [1]. Hence, there is a need to convert PLGA nanoparticle preparation toward scalable manufacturing technologies.

This study aims at comprehensively investigating the manufacturing performance of PLGA nanoparticles prepared by high shear mixing (HSM) followed by high pressure homogenization (HPH) or microfluidization (MF). After a standardized pre-emulsification step, a formulation and process optimization were conducted with both devices by means of Design of Experiments (DoE). This involved investigating the effect of input parameters such as stabilizer concentration, proportion of organic solvent, pressure and number of cycles on the resulting particle size and particle size distribution (i.e., polydispersity index (PdI)). The optimized formulations were further characterized regarding surface charge, stability and shape using cryo-scanning electron microscopy.

By means of DoE, it was found that for both HPH and MF the stabilizer concentration, the pressure, and the number of cycles exerted the most significant influence on particle size and PdI. Compared to HPH, manufacturing via MF resulted in smaller particle sizes and a narrower particle size distribution. These results can be attributed to lower pressure fluctuations due to the electronically controlled pressure pumps of the MF device in contrast to the manually controlled valve of the HPH device. Formulation and process optimization revealed particle sizes in the range of approximately 140 (MF) to 150 (HPH) nm with a narrow particle size distribution (PdI values below 0.1). Moreover, they exhibited a high physical stability, a negative surface charge and an almost spherical shape. This highlights that both technologies are suitable for the large-scale manufacturing of PLGA nanoparticles, whereby a more reliable performance can be expected with MF.

REFERENCES

[1] M. C. Operti, A. Bernhardt, S. Grimm, A. Engel, C. G. Figdor, and O. Tagit, "PLGA-based nanomedicines manufacturing: Technologies overview and challenges in industrial scale-up," *Int. J. Pharm.*, vol. 605, p. 120807, Aug. 2021, doi: 10.1016/J.IJPHARM.2021.120807.

[2] H. Havel *et al.*, "Nanomedicines: From Bench to Bedside and Beyond," *AAPS J.*, vol. 18, no. 6, pp. 1373–1378, Nov. 2016, doi: 10.1208/S12248-016-9961-7/TABLES/1.

[3] J. M. Metselaar and T. Lammers, "Challenges in nanomedicine clinical translation," *Drug Deliv. Transl. Res.*, vol. 10, no. 3, p. 721, Jun. 2020, doi: 10.1007/S13346-020-00740-5.

408e Improving the Stability of Nuclear Thermal Propulsion Fuel Elements with Tungsten Atomic Layer Deposition

Davis R. Conklin, Sarah Bull, Christopher G. McKinney, Jamelle K.P. Williams, Arne Croell, Jhonathan Rosales, Alan Weimer

Crewed missions to Mars will require engines with high specific impulse to achieve sufficient fuel efficiency for a return trip. Nuclear thermal propulsion (NTP) engines can achieve double the specific impulse of conventional chemical rockets by utilizing fission reactions to heat hydrogen propellant to around 2700 K before it is expelled through a supersonic nozzle. Fuel elements for NTP engines are typically composed of uranium oxide fuel pellets embedded in a refractory metal matrix and are susceptible to degradation at the temperatures required for NTP. Applying a barrier coating to the fuel pellets before fabrication of the cermet (ceramic-metallic) fuel element has been demonstrated to limit uranium hydride formation, prevent segregation of uranium oxide within the refractory metal matrix, and improve the final density of the fabricated cermet, all of which improve the durability of the fuel element. However, there is a need for finer control over the coating-fuel interface and an improved understanding of how coating properties influence performance in high-temperature, hydrogen-rich environments.

In this work, we utilize particle atomic layer deposition (ALD) to synthesize a tungsten coating on yttria-stabilized zirconia (YSZ) ceramic particles as a surrogate for UO_2. Particle ALD is performed by sequentially dosing WF_6 and Si_2H_6 into a fluidized bed of YSZ powder, which deposits a tungsten metal coating by self-limited surface reactions to create a nearly conformal 350 nm coating (Figure 1). Thermal treatments are used to evaluate the stability of the coated powders in a dilute hydrogen environment up to 1773 K. Further characterization of the powder by SEM-EDS after high-temperature hydrogen exposure is used to elucidate microstructural changes to the coating and fuel surrogate. To evaluate the ALD tungsten coating under more realistic NTP conditions, a test matrix is developed to compare the performance of the W-coated YSZ particles to the uncoated baseline YSZ particles in a variety of refractory metal matrices using spark-plasma sintering for consolidation. The consolidated cermet samples are then treated with pure hydrogen at >2000 K in MSFC's CFEET facility to simulate conditions approximating NTP engine operation. Microstructural analysis with SEM and XRD, as well as characterization of surrogate fuel mass loss is used to evaluate the effectiveness of the ALD tungsten coating in preventing cermet degradation. This is the first instance of an ALD coating being incorporated into an NTP cermet via SPS, as well as the first test of such a coating under realistic operating conditions. This research advances our understanding of how refractory metal ALD coatings behave under high-temperature and introduces a new technique to improve the durability of NTP engines for deep-space exploration.

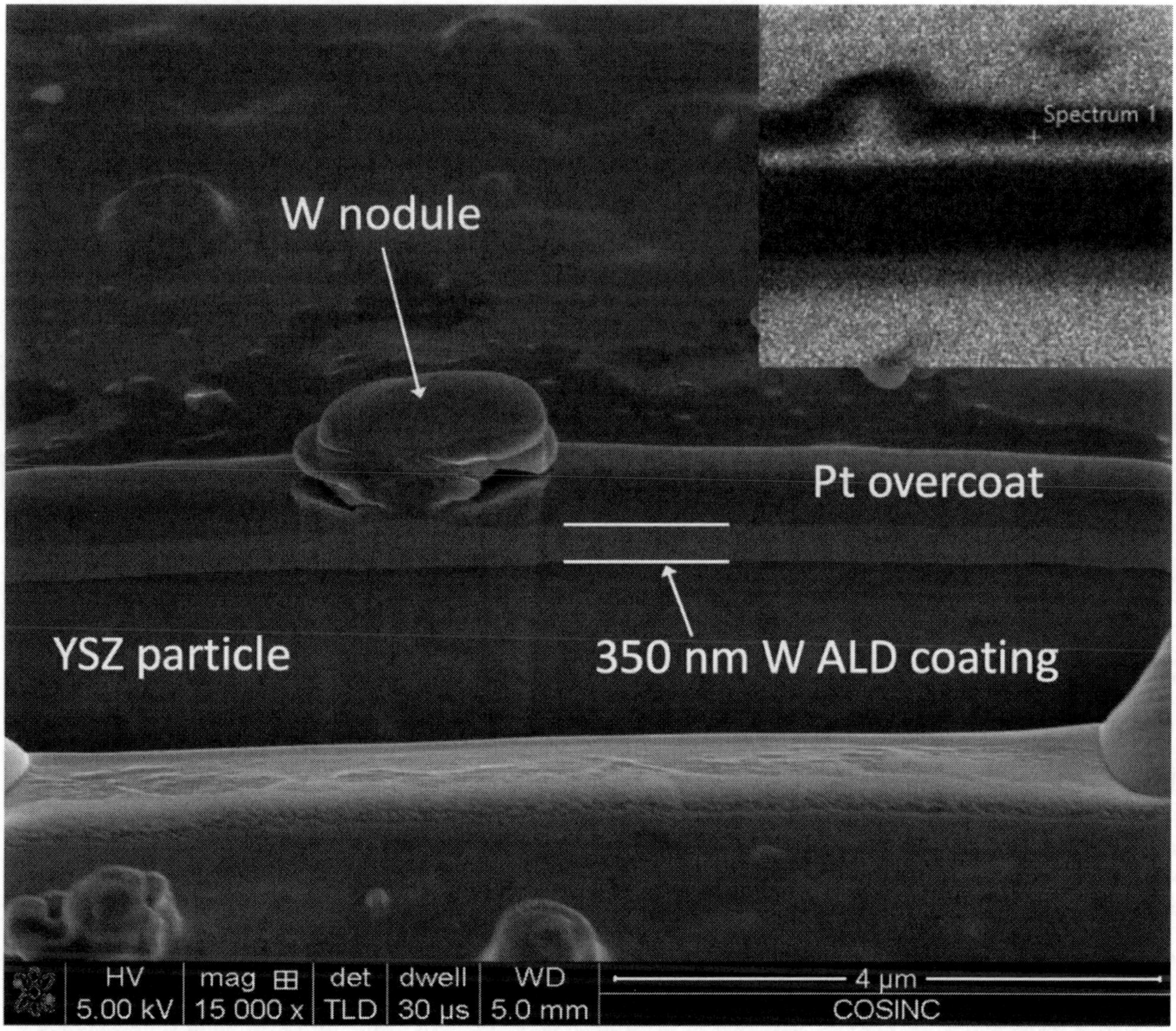

Figure 1: Cross-sectional SEM image of the ALD W coating exposed by FIB milling. Inset: EDS map of W Mα1 emission.

460b Enhancement of Crystallization Behavior for Solid Solution Separation Using Different Antisolvents

Vico Tenberg, Masoud Sadeghi, Andreas Seidel-Morgenstern, Heike Lorenz

Crystallization is utilized in a multitude of different applications in the chemical, pharmaceutical, and food industries, for example to directly purify target compounds or to crystallize and separate impurities. It can also be used to design a product regarding specifications like particle size, shape and particle size distribution, as well as application-oriented properties like flowability, compressibility etc.. Designing an efficient crystallization process to meet the desired specification and purity goals is challenging, in particular, if the target compounds cannot be purified in a single crystallization step. This might occur when highly purified products are required or if solid solutions are formed. A solid solution is a one-phasic solid, which consists of two or more compounds with a statistical distribution within the crystalline solid. As opposed to a co-crystal, which has a specific composition, solid solutions are present over a limited composition range as a partial solid solution or over the whole composition range as a complete solid solution.

The separation of solid solutions requires a multistage approach. Fractional crystallization is a classical method in which the resulting solid phase is redissolved and recrystallized for further purification. However, considerable amounts of the target product are lost in the liquid phase. An efficient process is counter-current crystallization (see Figure 1). Counter-current crystallization is to fractional crystallization what is rectification to distillation [1]. Due to stagewise remixing, one component is enriched in the solid phase, while another component is enriched in the liquid phase.

In addition to classical techniques like evaporative or cooling crystallization, alternative strategies such as antisolvent crystallization have shown potential to improve the efficiency of the counter-current process. Hence, distributions diagram can be constructed to compare different crystallization methods. For a typical binary solid solution, such diagram is obtained by plotting the solid composition versus the solvent-free liquid composition of one of the two components (see Figure 2). Similar to the well-known distribution diagram in rectification, it can graphically show the theoretical separation stages and help judging the separation efficiency for various conditions.

Recently, we proposed an isothermal process to separate an alyotrope found in the example system L-valine/L-leucine using water as a solvent and ethanol as an antisolvent [2]. Figure 2 shows the distribution diagram illustrating the alyotrope. Similar to azeotropes, alyotropes limit the purification at a specific composition depending on the process conditions. The process consists of a dual counter-current crystallization setup, where one cascade uses evaporative crystallization in water while the other uses antisolvent crystallization with ethanol as the antisolvent. The alyotropic point is shifted with changing process conditions similar to shifting the azeotrope via pressure change in a pressure swing rectification process.

This process was validated by predictive process simulation and experiments at a pilot plant scale. A more extensive process simulation of a five stage process (three evaporative and two antisolvent stages) at 25 °C with a feed composition of 25/75 wt.% L-leucine/L-valine supplied to the center crystallizer has been conducted. The results showed purities of 95.4 wt.% for L-leucine and 88.3 wt.% of L-valine with yields of 90.1 % and 90.8 %, respectively. The lower purification of L-valine as compared to L-leucine can be attributed to the overall low separation efficiency near the alyotrope (see Figure 2.).

Further work is underway to provide a more potent purification setup. We are going to consider different antisolvents at various process conditions. Moreover, thermodynamic-based models can be used to describe the distribution diagrams. The overall goal is to shift the alyotrope more efficiently and to improve the separation. The results of this investigation as well as extensive process modelling and related pilot plant experiments are to be presented in this contribution, which can be a basis to efficiently design a counter-current crystallization process.

[1] S. Münzberg, T. Giang Vu and A. Seidel-Morgenstern, 'Generalizing Countercurrent Processes: Distillation and Beyond,' *Chem. Ing. Tech.*, vol. 90, no. 11, pp. 1769 - 1781, 2018.

[2] V. Tenberg, M. Sadeghi, A. Seidel-Morgenstern and H. Lorenz, 'Bypassing thermodynamic limitations in the crystallization-based separation of solid solutions,' *Sep. Pur. Technol.*, vol. 283, p. 120169, 2022.

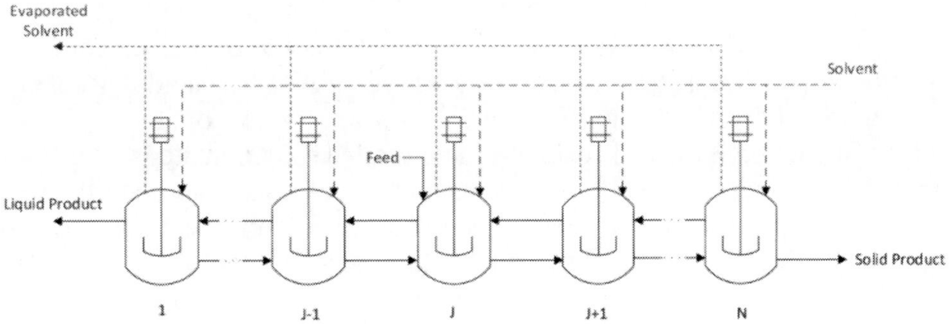

Figure 1: Simplified flow scheme of the counter-current crystallization process for evaporative crystallization. Solid lines: solid or liquid product streams; dashed lines: solvent used for dissolution; dotted lines: solvent evaporated in the process.

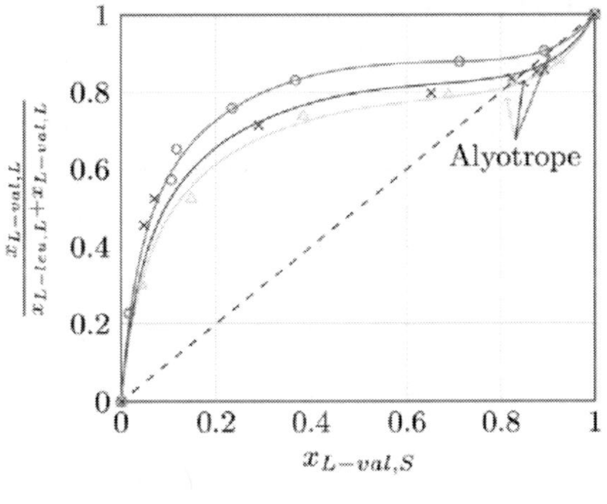

Figure 2: Distribution diagram of L-valine (*L-val*) and L-leucine (*L-leu*) dissolved in various water/ethanol mixtures at 25 °C. Experimental data points are described by an empirical polynomial model *[2]*.

460c Effect of Low Level Organic Impurities on Pharmaceutical Crystallization Kinetics and Impact on Physical Properties

Saurin Hiren Rawal, Jeremy Merritt

Abstract

The crystallization for current drug substance API manufacturing is usually designed to hit the major critical quality attributes, namely purity and physical properties. It has been well documented that impurities can alter crystallization kinetics to affect morphology, size and crystal form.[1-3] Many synthetic schemes have reactive crystallization for salt formation[4] as the final API isolation step where either a recrystallization is implemented to improve physical properties for downstream drug product processing. Herein, we present a case study where an API free acid is isolated for purity control followed by a final recrystallization for physical property improvement. We show how trace levels of impurities impact both the salt formation and recrystallization steps. We show how these impurities can directly influence the solid form landscape as well as deactivate the seed surface which leads to poor physical properties. We also show the efficiency of spherical agglomeration[5] to improve physical properties for downstream drug product development.

References

1. Mullin, J., *Crystallization, (Butterworth-Heinemann, Oxford, 2001).*
2. Rauls, M., et al., *The influence of impurities on crystallization kinetics–a case study on ammonium sulfate.* 2000. **213**(1-2): p. 116-128.
3. Ottens, M., et al., *Impurity effects on the crystallization kinetics of ampicillin.* 2004. **43**(24): p. 7932-7938.
4. McDonald, M.A., et al., *Reactive crystallization: a review.* 2021. **6**(3): p. 364-400.
5. Pitt, K., et al., *Particle design via spherical agglomeration: A critical review of controlling parameters, rate processes and modelling.* 2018. **326**: p. 327-343.

460f Continuous Crystallization in Slug Flow Aided with Indirect Focused Ultrasonication

Sunuk Kim, Consuelo Del Pilar Vega-Zambrano, Mingyao Mou, Frank Gupton, Mo Jiang

Many materials have been synthesized and manufactured to crystalline form for final products. [1–5] The crystallization process and crystal property are largely affected by compound-specific properties, such as nucleation tendency and crystal morphology. Ultrasonication technology has been useful in various industrial and academic studies, including reducing crystal size and aspect ratio, reducing nucleation induction time, narrowing metastable zone width and increasing yield.[1–5] This presentation talks about using indirect focused ultrasonication for facilitating nucleation, and improves the crystal morphology by combining sonication-aided nucleation with slug flow.

Reference

[1] Luque de Castro, M. D.; Priego-Capote, F. Ultrasound-Assisted Crystallization (Sonocrystallization). *Ultrason. Sonochem.* **2007**, *14* (6), 717–724.

[2] Sander, J. R. G.; Zeiger, B. W.; Suslick, K. S. Sonocrystallization and Sonofragmentation. *Ultrason. Sonochem.* 2014, 21 (6), 1908–1915.

[3] Ruecroft, G.; Hipkiss, D.; Ly, T.; Maxted, N.; Cains, P. W. Sonocrystallization: The Use of Ultrasound for Improved Industrial Crystallization. *Org. Process Res. Dev.* **2005**, *9* (6), 923–932.

[4] Lovette, M. A., andamp; Doherty, M. F. (2013). Needle-shaped crystals: causality and solvent selection guidance based on periodic bond chains. *Crystal growth andamp; design*, *13*(8), 3341-3352.

[5] Ruecroft, G., Hipkiss, D., Ly, T., Maxted, N., andamp; Cains, P. W. (2005). Sonocrystallization: the use of ultrasound for improved industrial crystallization. *Organic Process Research andamp; Development*, *9*(6), 923-932.

[6] Jiang, M., Papageorgiou, C.D., Waetzig, J., Hardy, A., Langston, M., andamp; Braatz, R.D. (2015) Indirect Ultrasonication in Continuous Slug-Flow Crystallization. *Crystal growth andamp; design*, *15*(5), 2486–2492.

271a From Mfix to Ccsi to Exa: A Retrospective on the Career of Madhava Syamlal

David Miller

Madhava (Syam) Syamlal began his career as a contractor for the National Energy Technology Laboratory (NETL), and in 2021 he retired as NETL's Senior Fellow for Computational Science and Engineering, the highest technical position within the laboratory. Through his remarkable career, Syam led the development of MFiX, an open-source CFD code for dense, reacting multiphase systems. Over the past 26 years, MFiX has advanced the state of the art to solve challenging problems beyond capabilities of existing tools, while also serving as a foundation for advances in many commercial CFD codes. In addition, Syam has led the development of new tools and approaches that cross disciplinary boundaries, such as the linking of CFD and process simulation through the Advanced Process Engineering Co-Simulation (APECS) project. In 2010, Syam led the formation of the Carbon Capture Simulation Initiative (CCSI), a unique collaboration among national labs, universities, and industry to establish, demonstrate, and deploy new computational approaches to accelerate the deployment of carbon capture systems by reducing technical risk. The legacy of that contribution continues in $CCSI^2$ (Carbon Capture Simulation for Industry Impact), which is actively supporting multiple technology developers at the pilot scale. Most recently Syam led NETL's activities within the Exascale Computing Project to develop a next generation version of MFiX to take full advantage of the next generation of supercomputing hardware coming online in 2022. This presentation will highlight these significant contributions of Madhava Syamlal throughout his career at NETL.

271b Don't get stuck: A DEM model for particle-particle cohesion using simple bulk experiments

Christine Hrenya

State-of-the-art, particle-based methods for assessing particle-particle cohesion are often impractical for industrial use, due to the associated time, costs, and expertise involved in AFM measurements and the like. In this work, we assess the use of a commercial rheometer to extract the force- and energy-based quantities needed to formulate a simple yet sufficient, physically-based, particle-particle cohesion model. The resulting model is validated quantitively with a different experimental system, namely one designed to the angle of repose. This system includes both brief and enduring contacts (i.e., energy-based and force-based), thereby providing a robust experimental test of the model. The resulting cohesion model is useful both directly in DEM simulations and indirectly for obtaining closures to population balances (continuum models).

271c Modeling and Numerical Simulation of Concentrated Solar EnergyStorage and utilization using Packed and Fluidized bed Systems

Zeyuan Gao, Javad Abbasian, Hamid Arastoopour

In this study a radiative heat transfer model is developed and a computational fluid dynamics (CFD) approach is used to simulate concentrated solar energy (CSE) absorption and energy recovery using both packed and bubbling fluidized beds containing silicon carbide (SiC) particles. Radiative heat transfer plays a very important role when the temperature is high, such as the temperature of a medium upon receiving radiative CSE. In this work, the radiative heat transfer was modeled as an effective conductivity based on the Breitbach and Barthels study. For conductive heat transfer, heat is transferred by the contact area between particles and gas near the contact surface. Here, we used the Zehner-Bauer-Schlünder (ZBS) conductivity model to calculate the effective thermal conductivity. For fluidized bed filtered two fluid model (continuity, momentum, energy for gas and particle phases and granular temperature equation) was considered and simulated using ANSYS Fluent 16.1. The result of our two-dimensional simulation agreed well with the literature data on temperature measurements of SiC particles in a packed bed and in a bubbling fluidized bed.

271d Some insights from validation studies of CFD-DEM modeling for gas-solids flows from lab to commercial scale

Tingwen Li

Particulate multiphase flows are widely encountered in many industrial processes. With the rapid development of high-performance computers, advanced computational algorithms, and multiphase flow models, computational fluid dynamic (CFD) modeling has become a valuable tool to understand the complex multiphase flows in various applications. Specifically, CFD coupled discrete element method (CFD-DEM) has gained traction in modeling gas-solids flows in recent years. In this presentation, we will discuss some lessons learnt from some of the early validation studies of CFD-DEM for gas-solids flows. The presentation will cover some fundamental studies of gas-solids flows at a small scale using MFIX-DEM including micro fluidized beds of Geldart A and B particles, mini circulating fluidized bed, and traveling fluidized bed. We will highlight some of the key inputs to the model for successful reproduction of experiments, for example, the critical role of wall effect for lab-scale setups. Finally, we would like to present our recent modeling study on the horizontal stirred bed reactor for polypropylene production, which includes systematic validation of CFD-DEM at different scales and our strategy to model the scale up to the commercial reactor.

271e An assessment of drag models for predicting fluidized beds with two-fluid and multiphase particle-in-cell frameworks

Casey LaMarche, Jia Wei Chew, Raymond Cocco

Simulations of fluidized beds in a 20 cm diameter column are performed using multiple computational frameworks, namely, Eulerian-Eulerian—i.e., two-fluid model (TFM)—and Eulerian-Lagrangian—i.e., multiphase particle-in-cell (MP-PIC)—and compared against experimental measurements. Eight different drag model formulations are examined to investigate the influence of drag model formulation on fluidized bed behavior: Beetstra et al. (2007), Tenneti et al. (2011), Tang et al. (2015), Wen and Yu (1966), Di Felice (1994), Gibilaro et al. (1985), Ergun (1952) and Wen-Yu (1966) blend, Yang et al. (2004), and Syamlal-O'Brien (1989). Experimental measurements and simulation predictions from all drag models are compared. The simulation accuracy is assessed by quantitively comparing axial pressure gradient and entrainment flux predictions to measurements. A discussion of the accuracy of drag model formulation in predicting pressure gradients and entrainment rates for the different computational frameworks is presented here. The most accurate drag model formulations are determined through error analysis.

271f Gas-Solid Flow Models Based on the Kinetic Theory of Granular Flows: What Have We Learned Since MFIX Version 1?

Rodney Fox

Collaborating with his PhD advisor, Dimitri Gidaspow, and coworkers at NETL, Madhava Syamlal published the first opensource code for simulating gas-solid flows based on the kinetic theory of granular flows (KTGF) in 1993. The MFIX code (https://mfix.netl.doe.gov/) is now in release 22.1 and has been used by over 7000 researchers worldwide. Moreover, its success provided the foundation for many commercial and academic codes using the same fundamental modeling approaches developed for the Eulerian-Eulerian version of MFIX.

Although the KTGF only provides conservation equations for the solid phase, it is especially useful for modeling moderately dense gas-solid flows such as fluidized beds and risers. This is because the gas phase serves as a source of kinetic energy through drag to keep the solid particles in a state of "rapid granular flow". As opposed to dilute flows, particles in moderately dense flows undergo sufficient collisions to allow for an Eulerian description that tracks only the mass, momentum and kinetic energy of the particle phase. Notwithstanding this relative simplicity, the numerical solution of the coupled gas-solid conservation equations is wrought with "hidden" difficulties that must be dealt with carefully to enable robust simulations. Indeed, as embodied in MFIX, Syam's contributions to the numerical algorithms are one of the keys to its success.

In this presentation, I will give an overview of my group's contributions to enhancing the capabilities of the KTGF models used in MFIX. Specifically, I will discuss the treatment of polydisperse particles (e.g., with a wide particle size distribution) and the extension of the particle-phase conservation equations to handle relatively dilute regions where collisions are infrequent relative to spatial transport. Using high-speed compressible gas-particle flows as an example, I will show results based on state-of-the-art hyperbolic flow solvers that illustrate the tremendous advancements achieved in this field since Syam's seminal work on MFIX release 1.

271g Modeling Multiphase Flow Reactor Systems with Gas–Particle Reactions

Charles E. A. Finney

The availability of functional open-source codes such as MFiX have enabled computational studies for design exploration of granular chemical reactor systems. Such reactors might involve particle evolution during processes such as chemical vapor deposition or pyrolysis and oxidation of solid fuel particles. In these systems, a range of physical submodels, particularly accounting for particle evolution, are required for accurate simulation of the complex multiscale, multiphysics processes. We highlight two such applications in this talk. The first application involves predicting dynamical patterns in chemical vapor deposition coaters where an objective is to achieve very uniform coating of particles with near-zero failure rates. Statistical analysis involves tracking the thermochemical time history of each particle or in using dynamical metrics to scale up reactor vessels to achieve similar particle-trajectory statistics. The second application is an autothermal biomass pyrolysis reactor in which a small amount of air oxidizes char to produce process heat inside of the reactor. Design exploration to balance stoichiometry, feed rates, and other process variables with product yields is enabled by coupling particle- and reactor-reactor scale models.

271h Particle transport during dry powder inhalation

Mostafa Sulaiman, Xiaoyu Liu, Sankaran Sundaresan

We have performed Computational Fluid Dynamics–Discrete Element Method (CFD–DEM) simulations of air and particles in a commercial ELLIPTA® inhaler [1]. We simulated the fluidization, deagglomeration, and transport of carrier and API particles, with inhalation profiles representative of moderate asthma and very severe COPD patients and different mouthpiece geometries and powder doses. In each simulation, we determined the fine particle fraction (FPF) in the stream leaving the mouthpiece, the temporal evolution of the spatial distribution of the particles, the mean air (slip) velocity seen by the carrier particles, and the average numbers and normal impact velocities of carrier-carrier and carrier-wall collisions inside the inhaler. In the cases examined, the air-carrier and carrier-carrier interactions affected the FPF, while the carrier-wall interactions were too infrequent to have a substantial effect. The simulations revealed the benefit of loading both blisters even when only a single medication is delivered.

We then performed integrated simulations considering both the inhaler and the Mouth-Throat (MT) region to probe the deagglomeration, transport, and deposition in the MT region [2].

The principal findings from these simulation studies will be summarized in this presentation.

References

[1] Mostafa Sulaiman, Xiaoyu Liu, and Sankaran Sundaresan, A CFD-DEM investigation of powder transport and aerosolization in ELLIPTA® dry powder inhaler, submitted for publication, 2022.

[2] Mostafa Sulaiman, Xiaoyu Liu, and Sankaran Sundaresan, under preparation, 2022.

AUTHOR INDEX

AIChE
120 Wall Street, FL 23
New York, NY 10005-4020

ISBN 978-1-7138-7916-9

CURRAN ASSOCIATES INC.
proceedings.com